GROWING UP WITH

SPACEFLIGHT:

PROJECT MERCURY

To BRADLEY THINK ROCKETS

Wes Oleszewski

8/17/19

To Roy Neal, the NBC reporter who was among those that brought the space program to us from the beginning.

Series edited by
Jim Banke and Pat McCarthy

Some people got to experience America's space program up-close and personal, hands-on, steeped in the excitement of the peaceful advancement of human civilization.

The rest of us had to watch it on TV.

"Everything that we do in our adult life stems from something that we went through in childhood that other people blew off."

- My Mom, November 22nd, 2014

GROWING UP WITH

SPACEFLIGHT:

PROJECT MERCURY

FREEDOM 7:
A HUMAN SACRIFICE – LIVE!

Although there were plenty of satellites being launched into space and getting mentioned in the news, for most Americans, the "space program" didn't really seem to begin until May 5, 1961, when the United States launched its first astronaut. The flight excited the American public like no other space event before, mainly because it could be seen live on national TV and heard live on the radio. This time there was a man on top of the rocket and many were sure they would see or hear him killed, right from their living rooms.

You see, the last live memorable space event to be broadcast nationwide took place on December 6, 1957, when America's Vanguard 1 rocket lifted about three feet from the pad and then crumbled back down into an explosive ball of fire. To the average slob on the street in 1961 there was little difference between Vanguard and Shepard's Redstone, or for that matter the Mercury Atlas 3 flight that also had blown up just 10 days earlier. A rocket at Cape Canaveral was a rocket at Cape Canaveral to most people, and the common misconception was that they usually blow up. Thus, the audiences were glued to their black and white TV sets, many waiting to witness a human sacrifice – live!

Although that "a rocket is a rocket" perception may have infected the public at large, it could not have been farther from the truth. The Mercury Redstone was the handiwork, largely, of the original members of Wernher von Braun's team of German rocket scientists. Coming to America by way of "Operation Paperclip," these were the same engineers who had developed the V-2 missile and, along with it, a new level of maturity to the art of rocket science. Their very "German" methods of meticulous testing and demand for absolute precision led to an upgraded cousin of the V-2 called the Redstone. Constructed for the U.S. Army, the Redstone was assigned to the same basic role as the V-2; that of a long range artillery piece. Von Braun, however, had an ulterior motive in mind as his team worked on the Redstone. It was the very same ulterior motive that had gotten him thrown into prison by the Nazis during the war and almost got him shot. That hidden agenda was nothing less than manned spaceflight.

Making its first flight from Cape Canaveral's Launch Complex 4 on August 20, 1953, the Redstone quickly developed a reputation as a reliable rocket that did not blow up all the time. In fact, of the first 11 test launches, only one Redstone, number four, blew up. Overall, prior to Shepard's flight, a total of 68 Redstones had been launched from the Cape and only 10 had failed – only two of which actually "blew up" on their own, while the others were either blown up by range safety or simply fell short of their objectives and dropped into the Atlantic Ocean. Compare that to vehicles such as the Atlas or the Thor which, at the time, were blowing up on nearly half of their launches.

At the same time, the Redstone was also serving on active duty with the Army and thus had many more successful launches to its credit. So, Shepard's ride was not nearly as dangerous as some thought at the time because he had a very reliable and well tested launch vehicle beneath him. Still, all of this was new to the free world and so, everyone was watching- including my grandma.

FREEDOM 7:
THIS IS IMPORTANT

On the day of FREEDOM 7's launch I had just recently turned four years old and really had little awareness of what was taking place at Cape Canaveral. My mom had taken my younger sister to the doctor that morning and I had been left at my maternal grandparents' house. Although I was easily distracted, grandma kept re-directing me to the TV set. "Watch this," she kept saying, "this is important."

(See Image 1, page 74.)

Looking at the tall white thing on the screen, I had no idea what it was or what it was going to do. Yet, grandma kept me watching. When liftoff finally happened she seemed a bit excited hooting, "Oh! Oh! There it goes!" I was completely indifferent. She then excused me to go back to playing whatever I had been playing. Oddly, however, that black and white TV image of Shepard's Mercury Redstone sitting on the pad and grandma's high-pitched exclamations when it flew remains one of my earliest childhood memories.

(OTHER MEMORIES) Mary; Titusville, Florida:

"When I was little there were always missiles flying from over on the Cape. You never knew when they were going to go, but suddenly someone in our neighborhood would start shouting, "Missile! Missile!

Missile!" and pointing toward the sky above the Cape. Sure enough there one would go and we'd all begin shouting, "Missile! Missile! Missile!" Sometimes you could really hear it and sometimes they blew up. After a long time you would hear the explosion- it sounded like thunder in the distance. We'd say, "No more missile" and wait for the next one. In the early 1960s it was like almost one going up nearly every day and sometimes at night."

(See Image 2, Page 75.)

Flying aboard his Mercury capsule, christened "FREEDOM 7," Alan Shepard made a sub-orbital hop down the Atlantic Missile Range. Officially the flight was designated Mercury Redstone Three or, more commonly, MR-3. His flight lasted just 15 minutes and 22 seconds. On the ground the news media on both radio and TV really did not quite know how to actually broadcast a spaceflight. After all, they were used to broadcasting political conventions and ball games and this was nothing like that. In fact, considering that the Soviets did their launches in secret, it was like nothing that had ever been broadcast before.

Shepard rode the Redstone for 142 seconds. At that point the engine cut off exactly as planned and the escape tower jettisoned itself simultaneously. Although Shepard heard the event, he did not see any part of it other than the green light that indicated tower jettison. Next he heard the clamp-ring that held the capsule to the booster blow apart and saw the green light that indicated the event. He also felt what he described as "a little kick in the tail" when the posigrade separation motors fired and separated the spacecraft from the booster. Now Alan Shepard was flying in space.

Automatically the Mercury spacecraft rotated around to what was known as the "blunt end forward" attitude needed for reentry. From that position Shepard exercised FREEDOM 7's manual controls in the pitch, roll, and yaw attitudes and reported periscope observations for all of two minutes. Then, at five minutes and 14 seconds into the flight it was time for the retro sequence. The retros fired on time and one minute later the retro package was jettisoned. Shepard heard what sounded like the pack letting go and saw some small bits of debris out the porthole, as well as part of one of the restraining straps, but he did not get an event light indicating the jettison. Instinctively he hit the manual "JETT RETRO" override switch and the green event light illuminated. After years of preparation to fly in space and months of preparation for this specific flight, all of which involved countless problems that could occur, this single little glitch with the retro package jettison light was the only thing on Shepard's flight that did not go as planned.

FREEDOM 7's reentry carried a high "G" load and Shepard momentarily soaked up 11 G's – 11 times the force of what a person experiences in normal gravity, or 1 G. The high G loading lasted only a few seconds, after which Shepard simply sat there and dropped into the atmosphere. At nine minutes and 38 seconds into the flight, FREEDOM 7's drogue parachute deployed and shortly thereafter the main chute deployed normally. FREEDOM 7 splashed down so close to the recovery helicopters that even before Shepard could get a good start at reading his instruments and switch positions as required during the post flight process, the recovery "helo" had already hooked onto the capsule. Concerned that the capsule may be a bit too low in the water, Shepard radioed up and asked the recovery helo to lift FREEDOM 7 a bit higher. As soon as that happened he took off his helmet, hung it on the hand-controller, reached over his right shoulder and cranked the lever that opened the hatch. He then sat in the hatch sill and grabbed the horse collar that was lowered from the helicopter. Alan Shepard, the first American to fly in space was safely landed aboard the carrier USS LAKE CHAMPLAIN, and thus brought the Free World to a standing ovation.

Since then a number of myths and legends have grown around Shepard's historic first flight. From a Hollywood movie that played VERY loosely with the facts, to poorly produced documentaries, to sanitized accounts published by NASA and "LIFE Magazine," to personal accounts drafted from memories with plenty of wear on them – it is often hard to drum out something close to the actual events. Since my personal account of FREEDOM 7 ranges no farther than my grandma's TV set, I figured it would be fun to denote a few of those odds and ends of the mission here.

One commonly held myth is that Shepard reported aspects of his condition and those of his capsule as being "A-ok." In fact, Shepard never used the term "A-ok" during the mission. The term "A-ok" was actually popularized by NASA Public Affairs Officer (PAO) John "Shorty" Powers who announced the flight. Yet the term "A-ok" was instantly integrated into the general public's spaceflight vocabulary. In the early 1960s my parents bought me a little blue jumpsuit with astro-insignias all over it. The one on the right chest was a round white logo that had a red "A-ok" on it.

(See Image 3, page 76.)

Many people think that during Shepard's flight, the public heard his voice calling out readings and making observations as it happened. In fact not a single word that Shepard spoke during the mission was allowed to be heard live by the public. Only those who were inside NASA working the mission and had a "need-to-know" were allowed to hear the live air-to-ground transmissions. This may seem amazing today when nearly every word of a spaceflight is broadcast live. Yet, although NASA was far less risk averse in the Mercury days, they were far more image aware. All of the air-to-ground transmissions that the Mercury astronauts broadcast were filtered through PAO Shorty Powers, who then repeated to the public a "NASA-correct" version of what was being said. Later, shortly after the flight, films of the mission had the edited recorded voice track of the astronaut dubbed in and from there on that became the "official version" of the mission. As the Gemini program came into being, the voices of the crew on orbit were allowed to be broadcast, but were normally done so by NASA on a time delay. Reentry and launch air-to-ground was not allowed to be heard live by the public until, believe it or not, Apollo 10. Prior to that, since the live broadcasts were quickly forgotten by the general public, those official films and clips that were later fed to the public soon developed the illusion that everything was always heard live. This became especially engrained in the public after Apollo 10, when all future launches and reentries carried live air-to-ground voice.

Exactly who would be the astronaut to fly aboard the first Mercury spacecraft was kept a close secret within NASA. All that was released to the public were the names of the top three choices: Glenn, Grissom and Shepard. It was not until after the scrub of Shepard's original launch date on May 2 that NASA decided to actually make public his identity as the first United States astronaut to fly. There were very elevated concerns that if the Soviets found out which of the seven Mercury astronauts were actually slated to make the first flight, the Soviets might assign agents to do harm to that individual. Historians digging in the KGB archives years later, however, have found no such inclination at all by the Soviets.

Shepard's FREEDOM 7 spacecraft was unique because it had a manually opened hatch. The hatch itself was closed with 70 bolts, but was released from the inside by way of a mechanical handle located over Shepard's right shoulder. Spacecraft 7 was the only manned Mercury capsule equipped with this hatch, which weighed 82.23 pounds. The other manned Mercury spacecraft all saved weight by having a pyrotechnically removed hatch, which weighed just 22.9 pounds. His spacecraft also differed from the other manned Mercury vehicles because it was equipped with two portholes rather than a forward-facing window. That forward facing window, however, ate up a large part of the weight saved in the use of the light weight pyrotechnically opened hatch.

(See Image 4, page 77.)

In that near-fictional movie titled "The Right Stuff," Shepard is shown stepping from the transfer van at the launch pad, stopping and looking up at the booster then giving a big "Thumbs Up." He did stop and look the booster over, but he never did the thumbs-up.

Additionally, the movie shows him stepping from the recovery helicopter onto the carrier deck and having his first footprints on the deck outlined in yellow paint. That never happened either. Of course those are just two of a large number of errors in that movie.

Shepard can be heard during the flight describing the "Beautiful view" seen through his periscope. That was a myth concocted by Shepard himself. As I stated earlier, FREEDOM 7 was the only manned Mercury spacecraft that did not have a "pilot's" window. Instead the spacecraft had two portholes located approximately where the astronaut's elbows would be. It was nearly impossible for a pressure-suited astronaut with a helmet on to see much of anything through those portholes. Like most of the manned Mercury spacecraft, however, Shepard's had a periscope. The periscope was a handy tool and the Mercury crews liked it. As Shepard waited through a series of pre-launch holds he had nothing to do other than look out through the periscope. At sunrise, as the vehicle sat on the pad, the sun shined directly into the periscope so Shepard put in a dark gray filter to cut down the glare. Unfortunately, when he remembered to remove the filter prior to launch, as he reached out for it the pressure gage on his wrist contacted the "Abort" button. Shepard immediately decided that he was not going to monkey with those filters anymore because initiating a pad abort, with the escape tower pulling the capsule away from a perfectly good booster because you were messing with periscope filters, would be hard to explain in the debriefing. During his flight, there was a specified point where he was supposed to look through the periscope and make a report. Dutifully, he deployed the periscope and discovered that the dark filter heavily obscured what he could see.

"I really couldn't see a damned thing through it," he confided to Gus Grissom after the flight, "so I just gave the same weather report that I'd been given in the preflight briefing and called out some known landmarks."

In fact, if you read the actual weather briefing given to Shepard before the launch and compare it to what he broadcast back to the ground it is almost word-for-word the same – with a few well known landmarks thrown in. In the transcript of his recorded debriefing, Shepard is evasive about this part of the flight and finally resorts to answering questions about his ability to see landmarks by simply saying "I don't remember" repeatedly. In that same debriefing he does, however, remember every other tiny detail about the flight. Hummmmm.

The service gantry used on Shepard's Mercury Redstone was a former oil derrick that was disassembled and trucked to Cape Canaveral during the early 1950s. It was set up at the newly established Launch Complex 3 & 4 to service the Army's first Redstone tests. The gantry was moved on common railroad rails in order to clear the rocket. Sometime between March of 1955 and April 1955 the tower was lowered and the gantry was moved to the Army's new Vertical Launch Facility. It was there that the gantry serviced FREEDOM 7's launch vehicle.

Shepard's launch gantry also had a special enclosure constructed on it that would surround the Mercury spacecraft and keep out the rain and the sun plus most of the sand and dust. The enclosure was air conditioned by a 10 ton machine and today would be called a "White Room." In 1961, however, there was a TV show that ran on ABC every Monday night starring Troy Donahue, Van Williams, Lee Patterson, Diane McBain and Margarita Sierra. The private eye type plots were set in Miami and the show was called "Surfside 6," so the folks at the pad decided to call their white room "Surfside 5."

It is also true that Shepard peed in his spacesuit while on the launch pad. The flight had been planned for just 15 minutes, so there would be just over four hours between the time Shepard was zipped into his space suit and zipped out of his space suit; and a good pilot can hold his water for that long. Unfortunately, assorted launch delays lengthened that time and the astronaut's bladder began to send signals insisting on being de-tanked. After some discussion, it was decided to cut the power to his bio-medical equipment and allow Shepard to wet his pants. On the next flight a sort of diaper arrangement was improvised.

So, although the flight of FREEDOM 7 was a historic event that was planned to the smallest detail, there were still some odd aspects to it. The legends and myths about the flight grew on their own, unfortunately aided by TV and movie producers who knew and cared nothing about spaceflight.

Although Soviet leader Nikita Khrushchev called the flight of FREEDOM 7 "a flea's jump," Shepard's flight was 100% successful. To many Americans it was a grand-slam home run scored in a game of spaceflight with the Soviets where we had previously been hitting foul balls and striking out. The flight helped influence President Kennedy to set the nation's course toward a landing on the moon. Additionally, in the wake of the Bay of Pigs political disaster the previous month, the FREEDOM 7 mission was the first major positive political event in the Kennedy administration. Overall, May 5, 1961, was a very good day in the free world... oddities, myths and legends aside.

BEFORE SHEPARD FLEW:
ASTRONAUTS- IN THE BEGINNING

Having been born in the spring of 1957 I was lucky enough to have quite literally grown up with spaceflight. Of course in those earliest years I was much more interested in things such as teething, stacking blocks and sloshing my hand around in the toilet- a skill that would later come in handy when I found myself living in a four bathroom home with a wife and two little daughters of my own. Once a month it seems as if one of the commodes is backing up and I need to slosh a plunger around. Still, America's manned space program took root in those first years of my life as well as the early years of the lives of many other folks.

During the late 1950s both the National Advisory Committee for Aeronautics (NACA) and the U.S. Air Force were looking into flying men in space. Meanwhile, ballistic missiles capable of delivering atomic bombs were rapidly being developed and pointed at each other by the Cold War adversaries of the United States and the Soviet Union. The Soviet warheads were far larger in dimensions and weight than those of the United States and thus they were pressed to develop a powerful booster to deliver them. The United States, however, had made breakthroughs in warhead miniaturization and thus required less powerful boosters for their delivery. The result set the stage for a series of Soviet firsts in space, which actually had less to do with war and much more to do with propaganda.

On October 4, 1957, less than half a dozen months after I was born, the Soviets placed Sputnik 1 into orbit around the Earth. Of course, I do not recall a thing about that event but the rest of the Free World was shocked. It was exactly the result that Sergei Korolev, the chief designer of the Soviet R-7 booster, had wanted. Now he had the political fuel to go to the powers that be in the Soviet government and say that by dominating space the Soviets could show the West that communism was better than capitalism. In return, he was directed to advance his work to accomplish even greater space achievements.

(OTHER MEMORIES) "Tommy"; Central Maryland: "When we heard that Sputnik was up it was like everyone got worried. I was in the 6th grade and my teacher, Mrs. Holder, seemed really worried. She told us that no one knew what that new Russian moon could do. I thought that it could go overhead during the night and drop an atom bomb on Washington, DC. My older brother kept looking up and once he spotted it and the whole family came running outside. There wasn't one, but two things in the sky zooming overhead like fast moving stars. I worried all night long that it was going to bomb us."

(AUTHOR'S NOTE: Often Sputnik spotters reported seeing two and sometimes three object passing overhead. It is often thought that these may have been the booster, the nosecone and the satellite itself.)

Indeed, there was a reaction in the United States

to Sputnik. President Eisenhower, who had always been indifferent toward any sort of space program, was suddenly faced with a public and Congressional outcry over the Soviet's accomplishment and soon was forced to act. The result would be the formation of the National Aeronautics and Space Administration, NASA, on October 1, 1958, just two days shy of the one-year anniversary of the launch of Sputnik 1. Although the new space agency absorbed the old NACA, and several other space- and rocket-related organizations, NASA was constrained by a very thin budget and very limited resources.

Using what they had, however, the new NASA began work on planning to explore space by way of machines and man. Out of that effort evolved "Project Mercury." Mercury consisted of a one-man capsule lofted into orbit by a rocket. Sounds simple, but that simple objective opened the door to thousands of challenges. What kind of rocket would be needed to loft this capsule? What would be needed to keep the man alive in space? Could a human even survive in space? Would the man survive the rocket ride needed to get into orbit? While he was in orbit, how would you communicate with him? Can you communicate from space to Earth in the first place? Once the man was in orbit, how would you get him back to Earth? What happens to simple materials during all of the phases of a spaceflight? All of these questions and countless more were unanswered during the late 1950s. Project Mercury started out with just a handful of engineers to answer those countless questions, and almost all of those engineers were aeronautical engineers and not spaceflight engineers; they would have to learn about spaceflight as they worked at solving the questions.

So lacking were the scientists and doctors in basic data about flight into space, that some really odd-ball concepts got real consideration- such as "Spudnik." Dr. Brown, A professor of biology at Northwestern University, had suggested to NASA that before committing an man to spaceflight perhaps it would be better to launch a piece of a common potato. The professor said that the potato has a "...very simple rhythm (of life), a 24 hour rhythm." And "It doesn't have a complicated life pattern." Thus, if a piece of potato were to be horribly killed by spaceflight, NASA would have to figure out why that simple life form failed in that environment. His students created a special little space capsule in which the hunk of potato could ride and that vehicle was dubbed "Spudnik 1." The idea made the evening news but, as far as my research shows, probably never made it aboard a launch vehicle.

One of the smaller questions that was commonly thought of as the largest problem, yet really was not, was what kind of man would occupy the Mercury capsule? The debate apparently was brief, as in November 1958 members of the powerful Space Task Group, (STG), which was the original cartel of Langley engineers charged with carrying out Project Mercury, came up with the qualifications they thought should be required to be an occupant of their capsule. Quoting from Dr. David Baker's book "The History of Manned Spaceflight."

"...applicants were to be exclusively male, between 25 and 40 years of age, less than 180 cm (70.9 inches) tall, in excellent physical condition, hold a degree in science or engineering, and at least three years experience in either physical, mathematical, biological or psychological sciences, three years experience in a field of engineering, or three years experience of aircraft, balloons or submarines, alternatively a PhD or six months medical work."

Those requirements sound very similar to the backgrounds of the members of the STG who composed them in first place, and apparently President Eisenhower thought so too. He rejected the requirements from the STG immediately and mandated that NASA must select candidates from graduates of military test pilot schools.

Although NASA retained the gender, age, height, physical and educational requirements during the first month of 1959, they now added that candidates must have at least 1,500 hours of flying time as a qualified jet pilot and be a product of a military test pilot school. The military was then handed the qualifications and managed the chore of coming up with qualified candidates in a very military manner. Unlike at least one Hollywood version of the selection process, where two NASA minions drive around the country seeking pilots, the reality was that it was the military who came up with records of 508 pilots who met the qualifications. NASA then trimmed down the list to 110 candidates, who were all sent orders to report to Washington, D.C., in secret, with no reason stated, during the first days of February.

Mercury astronaut Scott Carpenter stated in the book "We Seven" that the letter he received read something like this: "You will soon receive orders to OP-05 in Washington in connection with a special project. Please do not discuss the matter with anyone or speculate on the purpose of the orders, as any prior identification of yourself with the project might prejudice that project."

Thus, Carpenter went to Washington and onto a path that would lead to fame and glory beyond his imagination. Personally, at the time, I was a civilian who would not begin flying for another 18 years, plus I was only two years old, so I didn't get a letter ordering me to report to Washington... gosh darn it.

Once in Washington, the candidates for astronaut were briefed on Project Mercury. On his way to the briefing, Carpenter picked up the January 26, 1959, issue of "Time Magazine" to read on the plane. In its "Science" section was a story entitled "Capsule to Earth," which not only introduced the public to Project Mercury but also stated that 110 "stoic daredevil" test pilots were on their way to Washington to try out to be "the man in the satellite." So much for maintaining the secrecy. Carpenter said that when he read the article he thought, "Good Lord! That couldn't possibly be me!" He had been convinced that he was not qualified because although he had 2,800 hours of flying time, he only had 300 hours in jets and was convinced they were asking for 1,500 hours in jets. But the actual stipulation read "1,500 hours as a qualified jet pilot" Thus, in management-speak, once you are a qualified jet pilot, we want you to have 1,500 hours... of flying time, in anything. So, Carpenter was considered qualified to be invited to become an astronaut.

Following that series of initial "secret" briefings, each candidate was asked if he would be interested in the program. For those who said they would be interested in being the "man in the satellite," an additional series of tests and interviews followed. The group was then narrowed down to 32 "stoic daredevils." They would move on to the next phase of testing; the "medical" phase.

Located in Albuquerque, New Mexico, the Lovelace Medical Clinic was the site where the astronaut candidates were given seven-and-a-half days of medical testing that could easily be equated to torture. Most of this had little to do with flying and plenty to do with doctors gone wild. The testing protracted to the point where a couple of highly qualified candidates simply got fed up and left. One of those was Pete Conrad, who would later become an astronaut and fly four space missions as well as becoming the third human to step upon the Moon. On day seven of his torture at the clinic, Conrad got fed up and went into the clinic commander's office and dropped his sixth full enema bag on the guy's desk saying, "Here's something to remember me by," and then he returned to test flying. There were, however, 30 others who stayed through the entire process.

At length, NASA selected seven candidates to be the Mercury astronauts; Alan Shepard, Gus Grissom, John Glenn, Scott Carpenter, Wally Schirra, Gordon Cooper and Deke Slayton. They were introduced to the public at a press conference on April 9, 1959, just four months after NASA had set the firm qualifications for the candidates. America loved the Mercury Seven and the news media couldn't get enough of them, yet most people had no idea as to exactly what they were going to do or how they were going to do it – and neither did the astronauts. For the most part, Project Mercury was being designed as they went along. The controllers who came to Cape Canaveral to work the first Mercury missions were actually writing the procedures as they went along. Going into space with a man was something that was all new, and it was expected to move at high speed and on a slim budget.

BEFORE SHEPARD FLEW:
LITTLE JOE- NO, IT HAS NOTHING
TO DO WITH "BONANZA"

"LIFE Magazine" gained an exclusive contract to cover the lives of the astronauts by placing $500,000 in escrow for them and future astronauts to divide up. As a result, we in the American public got to see how they lived and trained by way of the pages of "LIFE Magazine." We were treated to photos of highly orchestrated family activities and well planed astronaut locker room scenes garnished with stories about the Mercury Seven. All the while, beyond "LIFE's" camera range, work was going on to address the challenges and concerns of doing what had never been done before – placing a human into space.

One of the earliest concerns in Project Mercury was "escape." Not escaping the Earth's gravity, but rather escaping an exploding booster. Considering that during the late 1950s it was not uncommon for boosters of the day to fail, often in a spectacular manner, it was decided that a way had to be developed in which an astronaut could get away from an exploding launch vehicle.

During the month of July, 1958, the Space Task Group's Dr. Max Faget (pronounced Fa-jay) suggested taking a solid propellant rocket mounted on a short tower and use it to pull the capsule away from danger. A solid fuel rocket would allow for a near instantaneous ignition and thrust great enough to pull a large, manned capsule away from danger. The idea was accepted and testing started. The problem was how to test this concept in flight without spending a lot of money. The space program was in its infancy and under the parenting of the Eisenhower administration, which had very little interest in manned spaceflight or spaceflight in general, so funds were scarce.

An answer to the question was to build a small booster that would use stock solid propellant rockets to boost the capsule and escape system along a flight profile that would mimic the actual manned booster. The escape system could then be flight tested. That small booster was the brain child of Paul E. Purser and Max Faget. This simplified design was 48 feet in overall height and weighed a maximum of 41,330 pounds. It was 6.66 feet in diameter and was powered by four solid fuel Pollux and four Recruit rockets clustered in its hull. The motor arrangement, when seen from behind, looked like a "4" on dice and it reminding Max Faget of a "Hard Four" in the game of craps; which is known as a "Little Joe." He thus named the program "Project Little Joe." Contrary to myth, the name had nothing to do with the character by that same name on the TV show "Bonanza." In fact, the project was named in mid 1958 and the TV show premiered more than a year later on September 12, 1959.

(See Image 5, page 78.)

Developing a thrust of 250,000 pounds, the Little Joe booster could lift a maximum payload of 3,942 pounds. On January 29, 1958, the Little Joe flight test program was put on paper, and was later updated on April 14, 1959. Primary objectives of the tests were to investigate flight dynamics, check drogue parachute operations, determine physiological effects of acceleration on a small primate and check the spacecraft's aerodynamic characteristics. North American Aviation was awarded the contract for the fabrication of the Little Joe vehicle. Animal payloads such as small primates and pigs were planned for some of the Little Joe test flights. The pigs were later deleted from the testing.

On September 2, 1959, the first attempt at a Little Joe launch failed when the escape rocket fired 30 minutes before the planned launch and carried the capsule out into the ocean leaving the booster on the pad. A wiring fault was blamed. On October 4, the same booster with a new spacecraft and a dummy escape rocket was successfully launched from Wallops Island, Virginia, where all of the Little Joe flights took place.

On November 4, another successful flight was launched. On November 8, 1960 the first of the series with a McDonnell production spacecraft was prepared for launch. Meanwhile, the Mercury astronauts – who were headquartered at Langley, just a short drive from Wallops – were taking note of the Little Joe activity. Alan Shepard suggested that a Little Joe flight should be made with him aboard at some point in the future, but NASA management said "No." Management was proven right on November 8 when Little Joe 5 lifted off. The launch was normal until 15.4 seconds after lift-off, at which time the escape rocket motor was prematurely ignited. The spacecraft did not detach from the launch vehicle and remained aboard until impact with the Atlantic Ocean, where both were destroyed. Had Shepard been aboard Little Joe 5 as he had requested, he would have taken that long ride to an inglorious doom. On March 18, 1961, the sixth Little Joe was launched and lifted off normally, but 19 seconds later the escape tower fired prematurely in a situation closely resembling the Little Joe 5 flight, marking this flight as a failure. Finally on April 28, 1961 – just 16 days after the Russians put the first man into space – Little Joe 5B was launched and although there was a hung start engine, the flight succeeded in certifying the launch escape system. That was a good thing, because Alan Shepard was scheduled to fly aboard a Redstone on May 2, 1961, just four days later.

BEFORE SHEPARD FLEW:
MR-BD- UNINTENDED CONSEQUENCES

When asked what kind of fuels it took to get the United States to the Moon, few average folks can give the correct answer. Those who are space-buffs or who actually are a part of the aerospace industry can usually name off the propellants such as LOX, RP1, LH2 and assorted hypergolic fuels. Although they are correct, they often leave out one critical fuel without which no human would ever have set foot on the lunar surface. The "fuel" that was required in the greatest quantity was political fuel.

On March 24, 1961, an event took place that, through completely unintended consequences, would open wide the political fuel valves. It was the flight monikered as "MR-BD." Completely unknown to those outside the program, yet destined to significantly affect the space history books, this single launch caused a good deal of animosity inside Project Mercury. Thus, it became an event that is often overlooked, and deliberately shunned by NASA itself because MR-BD caused the dominos of history to fall away from the favor of the United States and in the direction of the Soviet Union. People working on Project Mercury soon saw its effect as causing their immediate embarrassment in the press and in the eyes of the public. The result of the MR-BD colored the USA as being second in the "space race."

Because the MR-BD mission was injected into the flight schedule, Alan Shepard's flight was pushed back nearly two months. That allowed the Soviets to place Yuri Gagarin into orbit two weeks ahead of Shepard's flight and, in the eyes of the world, "win" the glory of having put the first man into space. To many at NASA, the MR-BD was seen as an unnecessary schedule slip that cost them the prize of being first. With the hindsight of a historian, however, MR-BD can be viewed very differently.

MR-BD was an acronym that stood for "Mercury Redstone – Booster Development." It was an extra flight that famed German rocket scientist Wernher von Braun and his team insisted must be placed into the schedule of the Mercury Redstone test flight series leading toward Alan Shepard's suborbital mission. In the original schedule, Shepard's MR-3 mission was supposed to launch during the third week of March. If the Soviets did not pull a red rabbit out of their hat Shepard would become the first man in space and thus "win" the space race for the United States. The problem was that all three of the previous unmanned Mercury Redstone launch attempts had run into a series of small problems with serious and, in two cases, embarrassing consequences.

When the first Mercury Redstone, MR-1, attempted to launch it had a mismatch in a two-pronged plug that pulled out of the base of the booster at the instant of liftoff. That plug was designed so that when it pulled out, as the booster began to lift from the launch ring, one prong, being one-half-inch shorter than the other, would disconnect first. On all previous Redstone's that had not been a problem because none of them had been wired the same as the manned version. In this case, the two prongs disconnected 20 milliseconds apart. Although that time span seems extremely brief to a human, in the area of electronics where electrons flow at the speed of light, 20 milliseconds is a very long time. The Mercury Redstone's automatic system sensed the difference and shut the engine down. With that, the booster simply set back down onto the launch ring. But the electronic brain was still working and it saw this as a normal shutdown, as if the vehicle was in flight, so it automatically jettisoned the escape tower! With the escape tower gone and no acceleration being sensed, the capsule's programming told it to go into recovery mode and it then deployed the main and reserve parachutes which popped out like corks. To say that this was embarrassing would be something of an understatement.

MR-1A was the next attempt and although it appeared to fly normally, later data analysis would show a critical problem. The carbon vanes that jutted into the exhaust flame to steer the Redstone showed an unexpected vibration. The frequency and the magnitude of that vibration grew very near to the predicted lifespan of the servo motors that moved the vanes. Loss of one of them could have easily resulted in the loss of vehicle control. It was determined that the vibration was caused simply by "...the lowered second bending frequency of the Mercury-Redstone booster-capsule configuration."

MR-2 was next and would carry Ham the chimp. At first the flight looked normal, but for reasons that were unapparent at the time, the booster was ascending too steeply and depleting its fuel at a higher than normal rate. This was caused by the thrust controller's servo control valve being stuck in the full open position. The result was that the propellant fully depleted at exactly 137 seconds into the burn. That was 5.5 seconds before schedule and .5 seconds prior to where the integrating accelerometer was set to arm. The Abort Sensing Implementation System (ASIS) sensed the anomaly and commanded an abort, firing the escape tower and pulling the Mercury spacecraft and Ham the chimp away from the booster. Instead of an expected maximum 12 G acceleration, Ham got hit with abort-level acceleration. He splashed down 137 miles farther downrange than planned and took about 17 G's in the process.

Although NASA officials later displayed a healthy Ham as proof of a successful flight, the fact was that this had been an abort, pure and simple. The flight was actually aborted during boosted flight by the triggering of the ASIS. The fact that the chimp had survived was actually immaterial. Had Shepard or any other of the seven Mercury astronauts been onboard that flight, the Soviets, as well as American critics of the program (some of the most vocal of whom were actually advising President Kennedy), would have been quick to point out that all NASA had done was boost the capsule to the point where the mission was aborted by the automatic system designed to save the astronaut's life.

In the wake of the growing laundry list of little failures, the von Braun team decided that another unmanned flight was required to test the fixes for the problems. Members of the Space Task Group, which if you will recall were the engineers charged with building the foundation of NASA's manned space efforts, chose to ignore the von Braun team's list of launch vehicle issues. The STG deemed the problems to be minor. Thus, the STG recommended to NASA headquarters that Shepard's MR-3 mission should "go" according to its original schedule and launch in March. Even decades later, former members of the STG hold to their original position. In his autobiography "Flight," former Mercury flight director Chris Kraft states

"The Germans were embarrassed by the Redstone's performance on MR-2, and by their failure to predict its fuel flow."

Of course he ignores the fact that a stuck servo control valve nullified any fuel flow predictions made by "...the Germans..." or anyone else. One thing he does get right, however, is his statement that the STG gang was "furious" when von Braun insisted on another test. Shepard himself recalled in his autobiography "Moon Shot" that the problem was a simple relay and nothing more, which is also far from being factual.

In fact, in their March 20, 1961, memorandum addressing the problems with the first three Mercury Redstone flights, the STG itself cited a total of nine different issues as presented by the von Braun team at NASA's Marshall Space Flight Center (MSFC) in Huntsville, Alabama. They were: 1) Rudder and carbon vane vibration, 2) Instrument compartment vibration, 3) Thrust controller, 4) H202 tank pressure regulator, 5) Cutoff arming timer, 6) Roll abort sensor,

7) H202 system controller, 8) Man-hole LOX leak, and 9) Velocity integrator. It is easy to see that the issue was a chain of malfunctions that caused the flight of MR-BD to be placed into the schedule. NASA Headquarters agreed. MR-BD would fly in late March of 1961.

So put off was the STG over this decision that they refused to allow the von Braun team to designate the flight as MR-2A or MR-3 and, again according to Kraft's book, they forced "the Germans" to use MR-BD as the moniker. They also refused to allocate an actual flight version of the Mercury capsule and denied the use of a live escape tower. Instead the MR-BD was topped with a used boiler-plate Mercury capsule from the Little Joe 1B mission and an inert escape tower. Nearly a half century later, Kraft still referred to the MR-BD as being "...von Braun's unnecessary Redstone test."

In aviation, or aerospace, the way that you prevent an accident or a catastrophe is when you see a chain of circumstances or failures that may lead to the accident; you break the chain. You recognize the related issues as they add up and correct the problems. In the case of MR-BD that is what the folks at MSFC did. The flight of MR-BD went off smoothly as every one of the fixes made by the von Braun team worked. Now the stage was set for the first manned Mercury Redstone flight.

(See Image 6, page 79.)

By delaying the schedule, the MR-BD inadvertently played directly into the hands of the Soviets. On April 12, 1961, they launched Gagarin's Vostok spacecraft and claimed the title of having put the first man into space. They had the illusion that this milestone would be a morale-buster for the United States and those people in the West. The Soviets, however, were badly mistaken. There is nothing Americans hate more than losing, and coming in second is considered as just being the first one to lose. America's new president, John F. Kennedy, had campaigned on the notion that the United States was falling behind the Soviets in many areas, especially spaceflight (and rocketry, due to the "missile gap" as he referred to it). It was an easy point to make considering the indifferent posture the Eisenhower administration took toward manned spaceflight. Yet, the shockwave that went through the American public due to the Vostok mission's success now gave Kennedy the perfect opportunity to fulfill his campaign rhetoric.

Across the United States the idea that "the Russians are beating us" became entrenched among the public at large. In Congress, members heard from their home districts as the "What are you going to do about it?" questions came from every corner. Kennedy now had the political fuel needed, not only to set the course of the United States toward the Moon, but also to gain the public inertia required to actually accomplish that goal over time.

Had the STG, Chris Kraft, Alan Shepard, et.al, gotten their way and had Shepard flown before Gagarin, Americans would likely have thrown up their collective hands and said "We Won! The space race is over!" And it may very well have been the end as the political fuel to go beyond Project Mercury would have quickly evaporated.

Thus, MR-BD, just by happenstance, was a very key link in the chain of events that led to United States astronauts actually getting to the Moon. It served to open the valve that allowed the political fuel to flow, and Vostok provided the spark that ignited Kennedy and propelled a nation. Although often missed as an easily overlooked footnote, the unintended consequences of that single mission would ensure that toddlers like me, and perhaps you, would have a space program to grow up with.

AFTER SHEPARD FLEW:
TO THE MOON, AMERICA

Throughout the 1950s and 1960s Jackie Gleason was one of the most popular television personalities in the United States. One of his regular skits was titled "The Honeymooners" and centered around the character of a city bus driver named Ralph Kramden , played by Gleason, and his wife Alice, played by Pert Kelton and later more famously by Audrey Meadows. Whenever Ralph had his fill of Alice's goading he would wind up his fist, punch it into the air and shout, "...to the moon Alice!" One of the few television seasons between 1952 and 1970 when Jackie Gleason did not have a show on the air was 1960-1961. In that same time period, Ralph Kramden may not have been sending Alice to the moon, but a newly-elected President was going to send America to the moon.

On Thursday, May, 25, 1961, President John F. Kennedy stood in front of a joint session of the United States Congress and openly challenged our nation to land a man on the moon and return him safely to the Earth within a decade. At that moment very little of what would be needed to get a human onto and back from the lunar surface existed. Engineers had not even agreed on the method by which to fly there. The United States had only one man-rated booster, the

Mercury Redstone, and it did not even have the power to place a spacecraft into orbit. NASA's administrator, James Webb, had only been in office for three months and six days when the lunar challenge was made and nearly every engineer who worked on Project Apollo had yet to be hired.

President Kennedy's speech contained more than just the challenge to go to the moon. In fact the address was titled the "Special Message to the Congress on Urgent National Needs." It covered areas such as economic and social progress at home and abroad, national "Self Defense," our military and intelligence shield, civil defense and so on. Overall, the speech had nine subjects and "Space" was the ninth. When Kennedy got there, he not only made the challenge, but he also detailed the cost and asked the Congress specifically for the needed funds. He also warned about the lead that the Soviets had in the spaceflight capability. Then he went beyond the objective of landing on the moon and pointed toward the future. He said in part:

"I believe we possess all the resources and talents necessary. But the facts of the matter are that we have never made the national decisions, or marshaled the national resources required for such leadership. We have never specified long-range goals on an urgent time schedule, or managed our resources and our time so as to insure their fulfillment."

"I believe that this nation should commit itself to achieving the goal, before this decade is out, of landing a man on the moon and returning him safely to the earth. No single space project in this period will be more impressive to mankind, or more important for the long-range exploration of space; and none will be so difficult or expensive to accomplish."

He then went on to say:

"Let it be clear – and this is a judgment which the Members of the Congress must finally make – let it be clear that I am asking the Congress and the country to accept a firm commitment to a new course of action, a course which will last for many years and carry very heavy costs: 531 million dollars in fiscal '62 – an estimated seven to nine billion dollars additional over the next five years. If we are to go only half way, or reduce our sights in the face of difficulty, in my judgment it would be better not to go at all.

"It is a most important decision that we make as a nation. But all of you have lived through the last four years and have seen the significance of space and the adventures in space, and no one can predict with certainty what the ultimate meaning will be of mastery of space."

"This decision demands a major national commitment of scientific and technical manpower, materiel and facilities, and the possibility of their diversion from other important activities where they are already thinly spread. It means a degree of dedication, organization and discipline which have not always characterized our research and development efforts. It means we cannot afford undue work stoppages, inflated costs of material or talent, wasteful interagency rivalries, or a high turnover of key personnel."

"New objectives and new money cannot solve these problems. They could in fact, aggravate them further--unless every scientist, every engineer, every serviceman, every technician, contractor, and civil servant gives his personal pledge that this nation will move forward, with the full speed of freedom, in the exciting adventure of space."

With this speech, JFK ignited moon fever in the United States. Engineering degrees began to be the thing to get and the nation began to build its pride around its space program. There was no talk of "can't" or "won't," and NASA management was decisive in the program's direction. Perhaps the only flaw was the rush to beat Kennedy's deadline. At its peak, NASA's funding was about 5% of the national budget. Estimates are that as many as 400,000 people worked in one way or another on what became the Apollo program. From the guys who poured the cement that made the foundations of the needed facilities to the women who sewed the space suits; thousands of contractors got a piece of JFK's challenge.

To the moon America! We would indeed go.

(OTHER Memories) Ray; from Ohio: "When school started in 1959 I was in the 8th grade and that year a new TV show came on called "Men into Space." I remember it was on every Wednesday evening and it quickly became my favorite show. I had to make sure all of my homework was done on Wednesday nights so I could sit and watch it. The show wasn't a documentary or anything, it was sort of science fiction. But, they said it took place in the very near future. It starred Colonel Ed McCauley and they were always trying to do neat space adventures like fly around the moon or land on the moon. Since it was so realistic, (AUTHOR'S NOTE: lot of actual early Atlas rocket footage was used and the space suits were either Navy or Air Force high altitude pressure suits.) it made you feel like we could get to the moon in just a couple of years." (AUTHOR'S NOTE: I had never heard of this TV show before Ray mentioned it. It ran just one season from September 30, 1959 to September 7, 1960 on CBS from 8:30 to 9:00 in the evening. It was never widely in re-runs where I was growing up with space-flight. Yet, nearly every episode can be found, as of this writing, on the YouTube website and although they are somewhat campy, they are really fun to watch.)

LIBERTY BELL 7:
AMERICA'S SECOND MAN INTO SPACE SHOOT

Just 20 days after the flight of FREEDOM 7, President John F. Kennedy challenged the nation to the task of "...landing a man on the moon and returning him safely to the earth." It was a goal that would soon electrify the people of the United States and send NASA into something of a tizzy. At first glance it seemed as if the U.S. had nothing more than 15 minutes and 28 seconds of manned flight in a spacecraft that was little more than a ballistic container hurled into a sub-orbital path by a rocket without the power needed to reach orbit. That situation would have to change in a very big way.

Additionally, the Soviets had already orbited a man and the feeling in the general public was that "we" were way "behind." Thus, the "space-race" was never actually declared, it simply came about out of our own notion that the Soviets were beating us at something. NASA, who it was perceived held all of America's cards in manned spaceflight, looked like a weak and somewhat inept team to go up against the Russian bear. In fact, the symbol of Project Mercury actually had greater recognition than the blue NASA "meatball" that had been on Shepard's space suit.

At the time of Kennedy's lunar challenge NASA's

plan was to have all seven Project Mercury astronauts each make one sub-orbital flight and then have each fly an orbital mission. Once the deadline of a man on the moon by the end of the decade was established, however, that schedule went directly down the toilet. The sub-orbital missions were eating up precious time and man-power in their planning and execution. So, nearly a month before Gus Grissom's Mercury Redstone-4 (MR-4) flight, Mercury management had already cancelled MR-6 and were calling into question John Glenn's MR-5 mission. Its fate, NASA said, depended on how well MR-4 succeeded.

On Friday, July 21, 1961 Gus Grissom waited aboard his Mercury capsule atop a Redstone booster as the countdown to his launch proceeded similar to the one that lofted Alan Shepard back in May. I was only a four-year-old at the time and, according to my mom's recollection, was not even awake as the TV networks broadcast coverage of Grissom's launch.

Grissom was fully trained to fly MR-4, the hardware was in place and the ground support was readied, so in spite of the schedule pressure the flight went ahead as planned. Although Shepard had flown aboard Mercury spacecraft number 7, Grissom's capsule was number 11 on the production line. A number of changes were added into spacecraft 11, which more closely represented the orbital version of the Mercury capsule. The most noticeable of these changes was a forward-looking window placed directly in front of the astronaut's face. The change with the most con-sequence, however, would turn out to be a new hatch that could be explosively removed.

(See Image 7, page 80.)

Although this hatch had been "tested" and demonstrated on mock-ups and capsule sections, MR-4 would be the first time the hatch would fly on the complete Mercury spacecraft and through a complete mission profile. The hatch itself could be "blown" open by two different means; a plunger button on the inside, or a "T" handle and lanyard on the outside. It is a common misconception that the hatch was held on with "explosive bolts" or bolts that were individually rigged with explosive cores. Rather, according to the Mercury spacecraft familiarization manual "the bolts are inserted through the entrance hatch sill, which incorporates the explosive charge, and threaded into space craft sill." The explosive charge was actually a long band of material sandwiched between the hatch and the hatchway sill. A magnesium gasket, with inlaid rubber, formed the hatch seal when the hatch was locked into position by "weakened" bolts that had a tiny hole drilled through them at the shoulder. Detonation of the explosive severed the bolt heads and released the hatch. In order to blow the hatch following splashdown, the astronaut removed an initiator cover, pulled the safety pin from the plunger and then pressed the plunger. Depressing the initiator plunger caused two spring-loaded firing pins to strike the explosive charge percussion initiators and detonate the explosive charge. The charge could also be detonated with the safety pin still in place. It took four pounds of pressure to set the charge off with the pin out and 40 pounds of pressure to set it off with the pin in place. The standard procedure at the time of Grissom's MR-4 flight was for the astronaut, after splashdown, to remove the cap and then remove the safety pin then push the plunger when it came time to blow the hatch. The second way to detonate the hatch explosive was by way of an external lanyard that was tied to a small "T" handle and stored in a small compartment at the lower end of the hatch itself. That handle and its 42-inch-long coiled lanyard were tucked behind a small external door held on by a single "screw." It was designed to allow external responders to quickly rescue an incapacitated astronaut.

Of course what is perhaps the most important fact about Gus Grissom's MR-4 flight that anyone reading this should know is that it was nothing like the way Hollywood portrayed the flight in that mostly fictional movie "The Right Stuff." All they seem to have gotten correct was that Grissom flew after Shepard. The truth about Grissom himself is self evident. He was selected to fly the second manned United States space mission. That selection was not done by drawing straws or tossing a coin. It was done through nearly two years of close personal scrutiny conducted by the highest levels of Project Mercury's management. All seven astronauts were carefully watched and their abilities, personalities, level of judgment, decision making and engineering aptitude, as well as their short-comings were noted and compiled. Their ranking was not a matter of popularity, but rather one of demonstrated ability. The top three were selected to make the first three Redstone flights. The order in which they were ranked were; 1) Shepard, 2) Grissom and 3) Glenn. You did not make that short list by being a "squirmin' hatch-blower," as was implied in the insulting Holly-wood version of Gus Grissom.

At 3:55 am Gus Grissom was loaded aboard his

Mercury capsule, which he had named "LIBERTY BELL 7." It was his third attempt to get the mission off the ground. The first attempt, on Tuesday, July 18, 1961, had been scrubbed the evening before it was scheduled to launch due to weather. The second time, which was scheduled for July 19, the count got down as far as T-10:30 and they scrubbed again due to the weather. Since the Redstone had been fully fueled, the booster would require a full 48 hours to drain, purge and dry out before another launch was attempted. Finally, on July 21, after a night of cloudy conditions, the morning weather started to clear as the heaviest of the clouds blew inland at sunrise. Up to the north of the Cape, toward Jacksonville, thunder heads had already formed in the sub-tropical Florida morning. Yet it appeared the forecasters had been right when they predicted thinning clouds just before launch time. Indeed, a thin overcast remained until just before liftoff. The launch vehicle was also on its best behavior as it was fueled. Procedures went smoothly through the countdown and all was ready for the flight of MR-4.

It is hard now for some who had yet to be born at the time of these early manned flights to really get a sense of how the public considered manned space-flight during most of the 1960s. The pompous "been there done that" attitude concerning space events projected a half century later by President Obama and his administration did not exist at all at that time; in fact the attitudes were exactly the opposite. During most of the 1960s, President Kennedy's challenge to reach for the moon had energized the public. It did not matter to Americans, or to most of the free world, that Shepard had successfully made an identical flight less than two months earlier. Grissom's sub-orbital mission electrified the public once again. Hundreds of thousands of people had flooded into the Cape Canaveral area. Hotels were filled, campgrounds were packed and on the morning of the launch the beaches were crowded with spectators. Nearly all of these launch-watchers had their eyes turned toward the Cape and their ears turned toward their radios. NBC, for example, was broadcasting by radio not only to the United States but also to the BBC, the British Commonwealth Network and the Armed Forces Network. Additionally, Voice of America carried the flight world-wide. Television also carried the event live on all three networks. At the Holiday Inn where the astronauts lived while training at the Cape, the staff were all wearing buttons that said "Good Luck Gus." Hotels along the beach and Highway A1A were set to close for business at T-5 minutes in the count-down so that their employees could go to the beach and see the launch. Nearly every person who had an I.D. for access to the Cape Canaveral Air Force Station was inside the base fence and positioned to get a good view. This event was one of those that simply stopped a nation in its tracks.

As with Shepard's flight, NASA would release all re-al-time information concerning the flight through the filter of its Public Affairs Officer, Col. John "Shorty" Powers. The purpose being to keep a tightly held lid on anything that may go wrong during the mission. On Grissom's flight, however, NASA would discover that when things do go wrong there is no way to keep the lid on. The news media of the day really did not know how to broadcast a spaceflight, but they were learning fast. When NBC came on with coverage of MR-4 they called it "The second United States man into space shoot." They had gathered their few correspondents who actually had a working knowledge of spaceflight and used each to buoy-up their anchormen.

LIBERTY BELL 7:
RIGHT THROUGH A BLUE PATCH

One of the greatest problems in broadcasting a spaceflight was that as soon as the booster was out of sight, there was nothing left to present to the viewing audience but Shorty Powers telling what NASA saw fit to release. Even on a 15 minute sub-orbital flight, that can get fairly bland. Of the three networks, NBC had gotten smart. Knowing that NASA would not allow cameras into Mercury Control during an actual flight, they were able to convince the agency to allow cameras in to film the practice countdown and practice flight that took place a few weeks earlier. They also had on file films of a recovery crew picking up a Little Joe Mercury capsule. These were edited and spliced together so that once the Redstone was out of camera range, NBC could broadcast the images of Mercury Control and the recovery as if it were being shown live. It was a fairly slick way to fill in the missing visuals. They had also recorded some controller voice tracks to be dubbed into the video.

Shorty Powers did the countdown for the public as some of the radio hosts, almost instinctively, counted down aloud in duet. Excluding Shorty Powers, you could hear the tension in the voices of the other broadcasters. These were hardened missile launch reporters who had been covering scores of launches over the years, yet the reality that once again there was a man aboard this vehicle played on their minds. Launch veterans such as Jay Barbree, Robert Lodge, Richard Bate and Merrill Mueller, who had seen almost as many vehicle failures as they had successes, knew deep down what was really taking place here. Above the Cape a single hole in the thin overcast suddenly opened and a beam of bright morning sunshine shown down directly onto MR-4. It was inspirational.

Upon the countdown reaching zero the Redstone ignited and simply lifted off. Merrill Mueller's experienced eye, just four seconds after liftoff, saw a difference between Shepard's Redstone and Grissom's.

"It's a faster rise than Shepard's rocket," he exclaimed breathlessly, "It's very true, it's very steady, it's almost twice as fast as Shepard's rocket was!"

In the background the ear-splitting scream of the Redstone nearly drowned out Mueller's voice.

"Very straight, very true, it looks beautiful! Glistening in the sunshine with a great fireball on its tail and its little black nose pointed straight up at the heavens! Right through a blue patch!" he gasped, "The only blue patch in a funnel of clouds that surrounds us on all four sides!"

Indeed MR-4 had found the only opening in the sky over the Cape and arced right through it.

Just over a minute and a half into the boost Grissom was riding along at an easily tolerated 2.5 G's. The Mercury astronauts had trained for as many as 12 G's. Grissom had been briefed by Shepard that as the Mercury Redstone went through Mach 1, the speed of sound, he could expect some heavy buffeting.

But, the aerodynamic fairings on MR-4 had been improved after Shepard's flight and as the vehicle exceeded Mach 1, Grissom felt very little buffeting. At two minutes and 22 seconds into the flight the engine of the Redstone suddenly shut down as scheduled. Although strapped tightly into his form-fitting couch, Grissom's innards were still subject to Newton's first law of motion: Every object in a state of uniform motion tends to remain in that state of motion, in a straight line, unless an external force is applied to it.

Thus, as the Redstone's engine stopped accelerating the mass of the vehicle, Gus' inner ear fluid wanted to keep going straight ahead and his brain interpreted that signal as him tumbling head over heels. The sensation lasted for just a moment and Grissom was never disoriented. At that same moment the launch escape tower automatically jettisoned. He later recalled that he could hear the rocket motors of the escape tower as they ignited, he could hear the bolts holding it to the capsule blow and he saw it arcing swiftly off to his right. He was able to see the bright red tower against the black sky as it continued way out into the distance. Ten seconds after shutdown, as he was watching the escape tower depart, the bolts that held together the clamp ring that fastened the capsule to the booster blew and the posigrade thrusters ignited to separate the spacecraft from the Redstone. He heard and felt both actions as they took place. LIBERTY BELL 7 was now on its own in space – at least for the next few minutes.

To the rest of the world, Gus Grissom and his Mercury Redstone vehicle seemed to have rocketed from Cape Canaveral and simply vanished into the sky. All that the audience had was the voice of Shorty Powers and the voices of the radio hosts to tell them what was going on. This was when NBC's clever filming of the practice mission and the men in Mission Control came in handy. The whole thing synced very well with the actual events. After just a few minutes of watching this simulation you got the impression that you were watching what was happening, only live.

Aboard LIBERTY BELL 7, Grissom observed the automatic turn-around of the vehicle. He then tested the control system in pitch, roll, and yaw; he attempted to make an Earth observation out through his window but saw only clouds. It was not until he was at the apogee of his flight that he finally saw land jutting out from under the clouds. What he saw down there he recognized instantly as Cape Canaveral. "I could even see the buildings," he later wrote. Once he oriented his eye on the Cape, he could then recognize West Palm Beach and Daytona. Then it was time to get back to work. Grissom had to configure his spacecraft for reentry. He placed the spacecraft into entry attitude and introduced a roll rate of two revolutions per minute in order to stabilize LIBERTY BELL 7 during its reentry.

When Grissom's reentry began, the listening world heard very little other than Shorty Powers repeating Gus' calls of his G-loads. Powers reported that "This is a calm, cool, collected, business-like pilot." The Public Affairs Officer dutifully repeated that Grissom reported "a G" and two seconds later Powers said, "He's reported 9 G's, 10…" there was a protracted silence. In fact Grissom received a high of 11.2 G's as his spacecraft battered its way back into the Earth's atmosphere. At this point, NBC dubbed in the pre-recorded voice track to their pre-recorded film of Mission Control, and the illusion was complete... for the moment.

Aboard the carrier USS RANDOLPH, pool reporter Charles Batcheldor was watching the splashdown and landing through powerful binoculars while talking in a continuous narration of what he saw. His communications link to the network was somewhat ratty, however, so they were cutting to him only briefly. It is important to note here that there was no live television from the carrier. In fact, live TV coverage of splashdown and recovery operations would not take place for another four and one half years when Gemini 6 splashed down in December of 1965. So, when Charles Batcheldor was speaking, a still picture of the carrier was all that was presented on the TV. Even Mercury Control did not have a televised feed from the carrier. Meanwhile Merrill Mueller was reading from his schedule of flight events and blindly announcing what was supposed to be happening. Unfortunately, he misread the document and was several minutes ahead of what was actually happening. His narrative had LIBERTY BELL 7 already on the water when in fact it was still more than 3,000 feet in the air and descending beneath its parachute. Splashdown itself was never actually reported to Mercury Control. Instead, Shorty Powers, simply listening to the cross-talk between the recovery helicopters and the carrier, reasoned correctly that LIBERTY BELL 7 was on the water.

The procedure for recovery was for the helicopter to come in close to the spacecraft, hook on the forward attach point and once it had the capsule secured, Grissom would blow the hatch and, according to accounts of the procedure, then "get out." This meant that he would get about half way out of the hatch while the helicopter lowered a second line with a "horse collar," which Grissom would grab and place under his arms. He would then be lifted into the helicopter and both the capsule and astronaut would be taken to the carrier. As the formation of recovery helicopters, call sign "Hunt Club," approached and the lead prepared to hook on to LIBERTY BELL 7, Grissom radioed that he was going to take a few minutes to write down his switch positions before he wanted them to hook on. When that word was relayed to Shorty Powers, he actually gave a brief laugh as he reported it to the public and added,

"...this sounds like typical Gus Grissom. He was not about to do anything until he was sure everything had been done and in a business-like manner."

This of course is quite contrary to the Hollywood "Right Stuff" portrayal of a squirming, nervous, panicky Gus Grissom. He was apparently not in any hurry whatsoever to get out of LIBERTY BELL 7.

Just a moment after Grissom told his recovery helicopter to hold off, he heard a loud bang as the hatch blew off. His moves were instinctive; he tossed his helmet out of the way, grabbed the hatch sill and pulled himself out as the capsule began to rapidly sink. Just a few minutes prior he had made several moves that ended up saving his life. First he had removed all of the harnesses that held him firmly into the form-fitting couch. Next he released his helmet from the neck-ring that held it to his suit. And most importantly, he rolled up the neck dam. This was a rubber turtle-neck fitting that was designed to keep water from filling his pressure suit. Wally Schirra, a Navy man, had advised Grissom to use the neck dam. (See Image 8, page 81.)

Schirra had learned that lesson from a Navy tragedy that had taken place on May 4, 1961, the day before Alan Shepard's flight. On that day in the Gulf of Mexico, the Navy carrier USS ANTIETAM had launched the Strato Lab High V balloon mission. Aboard was Commander Malcolm D. Ross and Lt. Commander Victor G. Prather and they went on to set a world's record for balloon altitude, reaching 113,740 feet. Instead of being pressurized, their gondola was open and thus the two aeronauts wore the Mark IV pressure suits. These suits were very similar to the Mercury suits, which were dubbed the Mark IVA; both models of this suit were produced by B. F. Goodrich. When the Strato Lab High V splashed down, everything was fine, the rescue helicopters were overhead, the carrier was only a mile away and both men sat there with their visors open waiting to be picked up. When Prather tried to grab the rescue helicopters horse collar, however, he slipped and fell into the water. The water then flooded into the faceplate opening of his suit. He quickly sank and drowned before rescue swimmers could reach him. For that reason Schirra had urged Grissom, an Air Force man, to heed the Navy's lesson and use the neck dam. Now, as Grissom unexpectedly found himself floundering in water that was more than three miles deep, Schirra's advice had paid off.

Although Wally Schirra had learned the lesson from Strato Lab High V, NASA management apparently had not. When you watch the films of the helicopter trying to save the capsule, and Grissom is thrashing in the water, one fact comes through crystal clear. There are no less than four helicopters hovering on the scene yet there is not one rescue swimmer. The fact is that no records as yet uncovered show that there were any rescue swimmers in any of those helicopters. None! Likewise, there is no flotation equipment other than a single small flotation device that appears to have fallen out of the capsule and is subsequently blown away by the helicopter's rotor-wash. There was, in fact, a full flotation collar "under development and testing" during the spring of 1961 according to the Project Mercury Quarterly Report for that time period. Why NASA did not have this device already qualified and in place before astronauts were actually splashing down is a question that should have been asked long before the flight of MR-4.

Just 20 seconds passed between the time Grissom appeared in the open hatchway and the time that the capsule was completely submerged. The helicopter, however, had managed to hook on to the Dacron strap that was looped atop the capsule. Grissom discovered that his suit was leaking and slowly beginning to fill with water, and thus he was slowly sinking. His helicopter, however, was preoccupied with recovering the capsule and as it hovered there in a tug of war with the Atlantic Ocean. There was not enough rotor clearance for a second helicopter to get to Grissom.

Meanwhile, as all of that was happening, those well-planned broadcasting efforts back at Cape Canaveral were rapidly going down the drain. Shorty Powers decided to just start reviewing what the controllers had been doing and he was saying "We are still standing by in the control center…" a lot. He also managed to talk about how high, how fast and how far Grissom had traveled. In Mercury Control everyone was completely unaware of the life and death struggle that was going on in the recovery area. In fact, some of the controllers were happily congratulating one another. On TV the pre-recorded video was showing a helicopter flying in with a Mercury capsule in tow, and anchor man Frank McGee was forced to explain on the air that this was all recorded earlier. NBC's Mueller was talking about chase planes and a little air show that Gordon Cooper was doing overhead. He even joked with Jay Barbree that Grissom, who "Loves fishing so much," may be after some blue water marlins out there. Suddenly, someone at NBC heard through the static that Charles Batcheldor was talking about Grissom being in the water. They quickly switched to his commentary.

"The astronaut Virgil Grissom is out of the capsule swimming in the water," Batcheldor reported in a long single breath, "the capsule itself is apparently sinking lower into the water and there is fear that it may sink completely below the surface of the water."

Millions of ears around the world were now trans-fixed on Charles Batcheldor and his narrative. At that moment he had the attention of more people than any other human being on Earth.

"Three helicopters are hovering very close to the astronaut and the capsule out there. They're making their attempts to get a cable to Virgil Grissom who is protected, of course he will float, there is no problem about that and they'll get him out."

Batcheldor, like many others, was under the misconception that space suits were good flotation suits as well. He apparently was unaware of, or had forgotten about, what had happened to Lt. Commander Victor G. Prather less than two months earlier and there was no way for him to know that Grissom's suit was taking on water.

"The concern at this time is with the capsule itself," Batcheldor nearly barked, "and it looks like Grissom is coming up now..."

Batcheldor continued to narrate the effort of Hunt Club 1, the Number 32 helicopter, to pick up LIBERTY BELL 7 and pull it from the water, but with so much water inside the capsule it was an even match with the helicopter's lifting capability.

"The helicopter is hovering over and holding the capsule up," he described, "The capsule is, well, sinking badly in the water and that's the reason that Virgil Grissom got out of the capsule."

Batcheldor went on to talk for a moment about where the helicopter would land on the carrier deck, then he paused.

"The capsule has dropped," he stated flatly, "the capsule has been dropped by the helicopter." He added "...and whether they can get it again, or not is going to be a major problem."

Although there are assorted reports as to what happened aboard the helicopter, the most accurate seems to be that as the recovery helicopter Hunt Club 1 was struggling to lift the waterlogged capsule, with co-pilot John Reinhard, who had attached the cable to LIBERTY BELL 7, pilot Jim Lewis suddenly got a "CHIP DETECT" light. The bright amber light was located directly in front of the pilot in the gonna-get-your-attention-no-matter-what, position. Anyone who has flown any sort of large aircraft will tell you one of the worst lights you can get is the "CHIP DETECT." In fact it normally is just a bit down the list of worst lights from the "FIRE" warning. The light itself says that the system has detected metal fragments within your engine oil. This is very bad and in some aircraft it requires that you immediately shut the engine down, because something inside your engine is coming apart in a very big way. In other aircraft it indicates that you have a certain amount of time before you can expect the engine to simply fail, but in all aircraft it means that your day has just gotten really bad. In the UH34D helicopter that Lewis was flying the "CHIP DETECT" tells the pilot that he has about five minutes of time to get the aircraft on the ground before the engine quits. Thus, once Lewis saw the "CHIP DETECT," he had no choice other than jettisoning the capsule and heading for the carrier.

Meanwhile helicopter Number 7, flown by pilot Phil Upschult and co-pilot George Cox, picked up Grissom and flew him safely to the carrier where he was whisked below decks into the medical bay by doctors. Grissom later stated that when he looked up and saw Cox's face he was highly relieved. In most of Grissom's practice water recoveries, Cox had been the man hoisting him.

Oddly, it turned out that the "CHIP DETECT" warning light on helicopter Number 32 that day was actually a false alarm. It was just a few bits of carbon contamination that had bridged the magnetic contacts and set off the warning. No pilot on earth, however, world have bet on that- not even a Marine. If Lewis had taken the time to pick up Gus and then fly toward the carrier there was the chance that the helicopter could have crashed in the sea on the way there. Or worse, it could have lost the engine while hovering above the deck of the carrier attempting to land and made a large fireball killing the crew, the astronaut and a lot of people on the deck. The best decision and the right decision was to get his aircraft out of the way and let the other helicopter take over and pick up Grissom.

There was a stunned silence when the networks cut away from Batcheldor aboard the RANDOLPH. Mueller began to speak, but for the most part said that due to the ratty voice signal all he really got from it was that the capsule was lost and Gus was not. Then Shorty Powers came on with the blunt, sanitized NASA version,

"I'll repeat what happened downrange in the pickup area," he said stoically, "we know that the helicopter attempted a pick-up of the Mercury spacecraft, we know that Virgil "Gus" Grissom got aboard the helicopter, we also know that some kind of malfunction occurred, the spacecraft was dropped in the ocean and sank."

He then concluded NASA's official coverage of MR-4 by quite flatly saying,

"This is John Powers in the Mercury Control Center signing off at this time."

And that was that; the ending of MR-4.

LIBERTY BELL 7:
WONDERING JUST WHERE THAT
MERCURY CAPSULE HAD SUNK

After Grissom's flight all other Mercury Redstone missions were cancelled. This was actually due in part to the overriding success of MR-4, said NASA. Yet another factor took place a little over two weeks after Grissom's MR-4 flight. On August 6, 1961, the Soviet Union launched Vostok 2 and this time they kept cosmonaut Gherman Titov in orbit for a full day. From a political point of view it would be very bad for NASA to keep doing sub-orbital hops while the Soviets were doing day-long-duration orbital missions. Also, NASA had learned as much as they could from the Redstone missions and more Redstone flights would not get them any closer to the moon any faster. Thus, Project Mercury would now focus on the Atlas, and orbital missions.

No one ever accused Gus Grissom of having done anything wrong. In fact it is clear that he followed every procedure to the absolute letter. No one ever implied that he was nervous or that he panicked – until the book and the movie "The Right Stuff." It was then that the true Gus Grissom was brushed aside, and a cartoon-like caricature of a brash, knuckle-dragging screw-up who happened to get picked as an astronaut was presented to the public. Wally Schirra described the movie as "Animal House in space." At best it is a work of fictionalized history, at worst it is a work of Hollywood character-assassination, especially where Gus Grissom is concerned.

So what really did happen to the hatch on MR-4? No conclusion has ever been made officially, but there is a chain of circumstances that can be considered in order for you, the reader, to come to a conclusion of your own. First, it was a new, un-flown modification to the spacecraft. Second, the procedures for uncovering and arming the hatch had never been conducted with an actual loaded hatch on an actual spacecraft by an astronaut in a pressure suit. Third, in later flights, every astronaut who actually hit the button to blow the hatch ended up with a deep welt on their hand because the plunger recoiled and made the wound right through their glove. After his flight Grissom had no such mark on any part of his body, including his hand. Fourth, with the safety pin pulled it took just four pounds of pressure to trigger the hatch. You can place an egg on a scale and push on it for four pounds of pressure and not break the egg. Fifth, Grissom was obviously in no hurry to get out of the spacecraft. Sixth, and perhaps most importantly, NASA management had provided zero safety support during the on-water portion of the recovery. No flotation devices were provided for the astronaut nor was a flotation collar provided for the capsule and there were no rescue swimmers on the scene. Frankly, that is so negligent and short-sighted I can hardly even write about it. All that combined led to the loss of LIBERTY BELL 7.

So far as Grissom himself is considered, if he

was the screw-up that Hollywood would lead us to believe, why then was he placed in command of the first Gemini mission (after Shepard was grounded) and the first Apollo mission? I'd say it was because he was the demonstrated best of the best and very highly regarded among the astronauts.

The most logical explanation of what may have happened to MR-4's hatch came, in my opinion, from famed Pad Leader Guenter Wendt, who was one of the people most familiar with the Mercury spacecraft and its escape systems in their full flight configuration. He placed the most probable blame upon the external handle and lanyard intended to be used to blow the hatch from the outside. His thought was that the door covering the device came open and the handle and lanyard dropped out and became entangled in the capsule's landing bag straps. As the capsule rolled in the waves it gave that four pound tug on the lanyard and blew the hatch.

Considering that I was just a couple of months into my fourth year of life and that my mom says I probably had not awakened for the day by the time the flight was over, I probably came shuffling out of my room, dragging my favorite stuffed monkey by its tail, rubbing my eyes and asked, "What's this?" when I looked at the family TV. To which my mom probably answered "I don't know." The fact is, when the coverage ended no one really knew what happened, and today we still do not know for sure. I do recall the grown-ups talking about how the Mercury spaceship had sunk in the ocean.

A few weeks after the flight, my folks rented a cottage on Whitestone Point on the Michigan shore of Lake

Huron. To a four-year-old growing up in Michigan, the lake simply is the ocean and I recall looking out across that expanse of blue water and wondering just where that Mercury capsule had sunk.

On the first day of May, 1999 an expedition headed by famed deep ocean salvage expert Curt Newport located and recovered LIBERTY BELL 7 from 16,000 feet of water in the Atlantic Ocean. His team, funded in part by The Discovery Channel, recovered the capsule and working with the Kansas Cosmosphere restored the vehicle for its display there.

THE ATLAS:
A BOOSTER THAT BLOWS UP
AS OFTEN AS IT GOES UP

On the 12th day of April 1961, deep behind the Iron Curtain of the Soviet Union and under absolute secrecy, cosmonaut Yuri Gagarin was launched into orbit. After circling the Earth one time his Vostok spacecraft re-entered the Earth's atmosphere. Somewhere around 23,000 feet Gagarin blew the hatch, ejected from the spacecraft and parachuted to the ground. He thus became the first human to orbit the Earth. At that exact moment in time, the schedule for astronauts of United States to fly in space consisted of a series of seven manned Mercury Redstone suborbital flights followed by an equal number of manned Mercury Atlas orbital flights.

Exactly where the Soviets were in their progress toward achieving spaceflight was pretty much unknown to everyone in the West prior to Gagarin's flight. Indeed, NASA's manned space efforts were trudging along lethargically, constrained by politics and a meager budget that had been allotted to it by the Eisenhower administration. When the success of Gagarin's spaceflight was revealed to the West everything changed.

Similar to the situation our nation would find itself in a half century later, on that April day in 1961,

NASA found itself without the capability to place U.S. astronauts into orbit aboard any launch vehicles in the United States inventory, while Russia was happily and successfully able to place people into orbit atop their reliable R7 booster. The difference between then and 2012 being that not only were the American people actually paying attention to that shortcoming, but the President of the United States himself was actually willing to lead the nation toward rectifying the situation. Less than one month following Gagarin's flight, President Kennedy challenged the nation and the Soviet Union to a space race. Although 50 years later, in era of President Barack Obama, it seems to be politically acceptable for the president to offer the premise that the United States can "lead from behind," such was not the case in 1961. It was the era of the "Cold War" between the United States and the Soviet Union and JFK was not about to allow the Soviets to win this skirmish. Likewise, the United States Congress unified behind the president and provided the funds that he requested to energize America's space program.

Immediately the schedule of flights for United States astronauts was drastically revised. The suborbital, manned Mercury Redstone flights were scaled back from seven flights to only two flights. Following that second manned Mercury Redstone flight the effort was to be directed toward manned orbital flights atop the Atlas booster. For most of the previous two years the Atlas had been test flying Mercury capsules. In the scheduled flight rotation of astronauts the third man up, and the first human to ride an Atlas was slated to be Marine Corps Lt. Col. John Glenn.

Unfortunately, the Atlas-type booster that Glenn

would ride had seen a troubled childhood. In 1958, six out of 14 launches failed and 1959 was not much better as ten out of 23 failed. In that same year, from January to June, a total of eight vehicles in a row failed and the most embarrassing had to be number 7. On May 18, Atlas 7D was launched with the brand new Mercury 7 astronauts there as spectators. The vehicle exploded almost directly overhead as range safety hit the "Command Destruct" button. During 1960 the Atlas was still batting about .500 as 14 out of 33 launches failed, not counting Atlas-Agena vehicle failures. In 1961, just one year before manned flights on the Atlas were scheduled, 13 out of 37 Atlas launches failed. Granted, most of the failures were "R&D" launches, but a bad reputation still hung over the Atlas.

Yet the space race was fresh off the starting line and Mercury was America's only bet to put a man into orbit, and Atlas was the only launch vehicle in the United States inventory that was capable of doing that. The Atlas itself was unique because its skin was so thin it could not support its own weight. This meant that the vehicle was extremely light weight, but it also had to be pressurized in order to keep from collapsing. Interestingly as well was the fact that the Atlas was not a two stage rocket, rather it was a single stage rocket that dropped two of its three engines at about two minutes and 10 seconds after launch. The center "sustainer" engine would continue to burn until the five minute point after liftoff and then shut down. Two small vernier engines that had done some of the steering during powered flight would continue to burn for a short time and correct trajectory errors if needed. Of course for manned flight a Launch Escape System

(LES) was added along with the Mercury spacecraft, and the "Mercury Atlas" was pressed into service.

The first of the Mercury Atlas launches took place on September 9, 1959, and consisted of the Atlas D booster with a boilerplate Mercury capsule minus the escape tower. Planned to send the capsule on a ballistic, sub-orbital path, the test was dubbed "Big Joe." (See Image 9, page 82.)

At booster engine cut-off (BECO) the two dead engines failed to drop off of the Big Joe Atlas and then at sustainer engine cut-off (SECO) the capsule failed to separate. After several minutes and depletion of all the Reaction Control System (RCS) fuel, the capsule came loose and dropped inertly into the atmosphere. Because the Mercury capsules were designed with an off-set center of gravity that would allow them to passively seek a blunt-end-forward attitude as it contacted the atmosphere, the Big Joe capsule survived reentry and was actually recovered.

Nearly 11 months after Big Joe, on July 29, 1960, the test was repeated with a flight dubbed "MA-1." On that flight the Atlas exploded while trying to push the Mercury capsule's mass through the area of maximum aerodynamic pressure (Max-Q). It was determined that the upper section of the booster was too weak to support the capsule under the Max-Q dynamic load. The booster's adapter section simply crumbled into the lower fuel tank. Strengthening a small area of the Atlas adapter was the proposed cure to get the vehicle through Max-Q and MA-2 would test that fix. A full Mercury spacecraft was loaded onto Atlas 67-D and launched on March 21, 1961. That time the sub-orbital flight was a complete success. Oddly, the "fix" to the adapter was only needed for MA-2 because Atlas 67-D was the last of the all "thin skinned" boosters to be used in Mercury; MA-3 would fly with Atlas 100-D, which sported a thicker skinned version of the booster.

MA-3 turned out to be a successful failure as the booster lifted off on April 25, 1961, and went straight up... and just kept going straight up when it should have pitched over. Again, Range Safety hit the command destruct and the Atlas blew up directly overhead. This time, however, the capsule had an active Launch Escape System tower and the abort sequence activated itself when it sensed the booster's destruct. The escape system pulled the capsule clear in an unplanned but successful test of the vehicle's abort system. MA-4 was to place a full Mercury spacecraft into orbit with an electronic "astronaut simulator" aboard. It was a once-around mission that was completely successful; the date was September 13, 1961. For the first time a Mercury spacecraft was actually placed into Earth orbit. Two months later, on November 11, 1961, Enos the chimp rode MA-5 on a successful flight. NASA then felt that the time had come to launch astronauts on the Atlas booster.

(OTHER Memories) Hugh; Cocoa, Florida: "I was pretty little when they were flying those Mercury flights, but I remember that we saw a lot of the rockets that launched from the Cape blow up. It seemed like the Atlas was the most spectacular- it was like a small atom bomb or something going off. We'd see one blow up and my dad would say, "Probably an Atlas." He could tell just by the fireball."

All of the Mercury Atlas launchings took place at Cape Canaveral's Launch Complex 14. So it was there that on February 20, 1962, John Glenn boarded his FRIENDSHIP 7 spacecraft that sat atop a booster many thought "blows up as often as it goes up." Glenn, like most folks inside the program, had a different perspective. He knew that each individual booster and each individual launch was its own individual event, and every failure had been a page added to the learning manual that improved the system as well as his chances of a successful mission. As the year 1962 started, John Glenn was ready to go... but it would take nearly two months for him to get to his three-orbit ride into history.

FREINDSHIP 7:
AGONIZING DELAYS

By the end of 1961 the concept of astronauts flying from Cape Canaveral had infected the American population and toy manufacturers were doing their best to cash in on it. Metal wind-up toys from Japan that had once been shaped like flying saucers were now arriving in the United States shaped like Mercury capsules. Toy maker Marx came out with the "Cape Canaveral Atomic Missile Base play set." The 1961 set included a lithographed steel headquarters building and fences. It also had spring-loaded rocket launching equipment that included a three stage rocket with a gantry for firing rockets such as Nike Missiles. Also included were BOMARC, Redstone and other famous rockets. Additionally, the set came with a searchlight, telescopes and a bag of plastic launch vehicle technicians. Of course it also contained a certificate of authenticity and a 32-page booklet of Cape Canaveral history plus a phonograph record. Shortly after the original run, a bag of plastic astronauts wearing Mercury Goodrich spacesuits was added to the set. Throughout the 1960s you could buy individual bags of those plastic astronauts, which for many of us became our very first space toy.

On the high end of the toy spectrum, model railroad

maker Lionel came out with their "Cape Canaveral Special" in 1961, which consisted of a cherry picker car with a silver suited Mercury astronaut in the cab. That car was set between two other flatcars; one with a rocket that actually spring-launched and had a capsule that came down on a parachute, and another that had two Mercury capsules being transported. Coming out in 1962 was the Marx "Space Ranger Base." This was a clever set that had a central "base" pedestal and a remote control that was connected by wires. From the pedestal extended a steel boom that hooked up to a rocket-shaped lander with a plastic propeller seated in it. Battery power allowed the user to make the lander lift off and fly in a circle around the base. It could also hover and pick up spacemen and other objects in the set. Of course, that was way out of my league. I had to settle for the bag full of blue plastic astronauts that cost 39 cents.

One of the cheapest and coolest little space toys was a mass-produced, hand-held plastic bottle with an accordion-like plunger on the bottom. The upper part of this was molded to resemble a Redstone booster and inserted into the upper opening of the bottle was a hard plastic Mercury capsule. Holding the bottle, you took the flat of your hand and hit the plunger- bottom and the capsule popped out into the sky. The side of the bottle had the word "Redstone" molded into it and the thing normally sold for just 29 cents. Of course as fast as the toy makers were designing and turning out these fun space items, they were going to have to work exponentially harder, because John Glenn was about to fly and his mission would excite the world even more.

John Glenn's mission to become the first American to orbit the Earth suffered an agonizing amount of delays. Initially planned to launch in the second week of January 1962, Glenn's MA-6 launch vehicle had suffered a fuel tank issue and was scrubbed. Thus, a new launch date of January 23 had to be set. On that Tuesday morning everything was ready, an estimated 600 newsman had gathered at Cape Canaveral, and around the world recovery and tracking stations were ready to go. The weather, however, consisted of heavy cloud cover and some developing thunderstorms. For the second time the launch of FRIENDSHIP 7 was scrubbed. The ill weather persisted for the remainder of the week. Meteorologists at the Cape forecast that Saturday, January 27, may provide good launch conditions. Once again John Glenn was suited up, loaded aboard his FRIENDSHIP 7 capsule and prepared for his launch. For a little more than five hours he waited as the count went down to the T-29 minute mark, when once again the launch was scrubbed due to weather. Mercury operations director Walt Williams expressed relief the launch had been scrubbed because of the weather, saying that "... Nothing was wrong, but nothing was right either."

In the process of de-fueling and inspecting the MA-6 booster following the scrub, technicians discovered a leak in the kerosene propellant tank. Repair of the leak would require a removal and reinstallation of the insulation around the liquid oxygen tank and the estimated delay would push the launch back to at least February 13. That date was subsequently pushed back to February 16 when, again, the launch would be scrubbed because of bad weather. Finally the weather conditions appeared favorable for launch on February 20, 1962. It was a date that John Glenn and FRIEND-SHIP 7 could finally keep.

FRIENDSHIP 7:
GOD SPEED JOHN GLENN

Launches of this era were far different than the launches that we came to know during the Space Shuttle era. Although long-range cameras showed the launch it was NASA's policy that the only thing about the launch that would be heard would come, as usual, through PAO Shorty Powers. He was the censor who decided what information was appropriate and what was not. So, just like the previous Mercury launches, although thousands of spectators gathered around the perimeter of Cape Canaveral, they didn't hear the actual countdown. Instead they heard Shorty Powers count backward to the launch. Everyone simply hung on every word spoken by him.

Awakened at 2:20 a.m. John Glenn proceeded through the now traditional steps of breakfast, medical exam by the "doctor of the Mercury 7" Bill Douglas and then the suit-up. Glenn's suit-up was tended to by suit technician Joe Schmitt who ended up suiting-up every United States astronaut crew from Freedom 7 to STS-4! At 5:16 a.m. he arrived with Glenn at launch Complex 14 on board the "transfer van," which actually was a semi-tractor-trailer truck.

(See Image 10, page 83.)

As the count entered a scheduled 90-minute hold,

technicians at the pad found a fault in the booster's guidance system. For a moment it seemed as if another scrubbed launch would be in the cards. But the problem was solved by the replacement of an electronics module and the delay in the launch was cut down to just 45 minutes. Then at 10 minutes after the hour of seven o'clock, technicians noticed a broken bolt among the 70 bolts that held the hatch closed. The bolt was ordered replaced and the launch was slipped by an additional 40 minutes. Finally at 8:05 a.m. the count was resumed. Two more delays came up in the count, one at 8:58 a.m., involving a fuel pump outlet valve, and the second at 9:25 a.m., involving a computer at the Bermuda tracking station. Each of these issues was rapidly overcome and the count continued.

Being a typical four-year-old boy of the early 1960s, I was busy bouncing around the tiny house that my parents had rented on Parkwood Street in Saginaw, Michigan. Although my mom had the TV on with John Glenn's Atlas shown in glorious black-and-white, I was far more interested in the half dozen lake perch that my dad had caught ice fishing on Monday night. As he caught them he set them, one-by-one, on the frozen lake surface next to his ice fishing hole. The result being that they had become frozen solid and remained so when he brought them home. We had no room in our tiny freezer for the fresh fish and if he had left them outdoors some critter would have likely found them and eaten them. So in order to keep them fresh my dad filled the bath tub with some water, dumped the frozen fish in and went off to work his normal third shift on the railroad. I woke up

Author's composite

As I remember it.
FREEDOM 7 on Grandma's TV

Image 1

Astronaut Alan Shepard

Image 2

NASA PAO, John "Shorty" Powers

Image 3

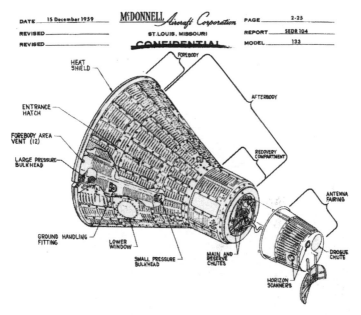

DATE 15 December 1959   McDONNELL *Aircraft Corporation*   PAGE   2-25

REVISED _____   ST. LOUIS, MISSOURI   REPORT   SEDR 104

REVISED _____   CONFIDENTIAL   MODEL   133

HEAT
SHIELD

FOREBODY

AFTERBODY

ENTRANCE
HATCH

FOREBODY AREA
VENT (12)

RECOVERY
COMPARTMENT

LARGE PRESSURE
BULKHEAD

ANTENNA
FAIRING

GROUND HANDLING
FITTING

LOWER
WINDOW

SMALL PRESSURE
BULKHEAD

MAIN AND
RESERVE
CHUTES

DROGUE
CHUTE

HORIZON
SCANNERS

Shepard's Mercury configuration

Image 4

Little Joe 5, November 8th, 1960.

Image 5

MR-BD lifts off, March 24th 1961

Image 6

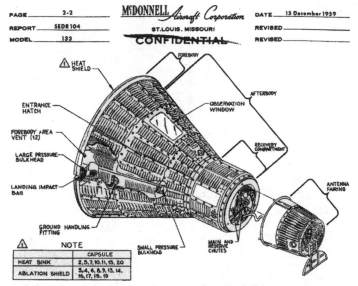

PAGE	2-2		DATE	15 December 1959
REPORT	SEDR 104	McDONNELL *Aircraft Corporation*	REVISED	
MODEL	133	ST. LOUIS, MISSOURI ~~CONFIDENTIAL~~	REVISED	

⚠ HEAT SHIELD

ENTRANCE HATCH

FOREBODY AREA VENT (12)

LARGE PRESSURE BULKHEAD

LANDING IMPACT BAG

GROUND HANDLING FITTING

FOREBODY

AFTERBODY

OBSERVATION WINDOW

RECOVERY COMPARTMENT

ANTENNA FAIRING

SMALL PRESSURE BULKHEAD

MAIN AND RESERVE CHUTES

⚠ NOTE

	CAPSULE
HEAT SINK	2, 5, 7, 10, 11, 15, 20
ABLATION SHIELD	3, 4, 6, 8, 9, 13, 14, 16, 17, 18, 19

Configuration of all Mercury capsules after FREEDOM 7

Image 7

Gus Grissom wearing the "neck dam" that saved his life.

Image 8

September 9th, 1959, Big Joe lifts off.

Image 9

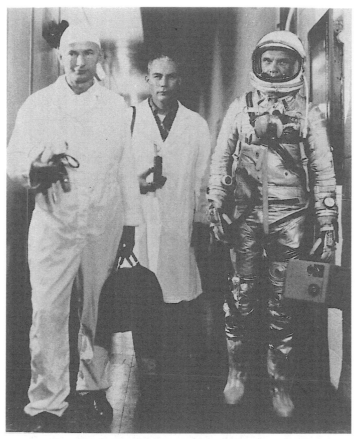

**John Glenn with Dr. Bill Douglas (center)
and suit technician Joe Schmitt (left).**

Image 10

FRIENDSHIP 7 lifts off at 9:47:43 am

Image 11

**Original 7 Mercury astronaut
Deke Slayton seen here in happier times.**

Image 12

**May 24th, 1962 Scott Carpenter
lifts off from Pad 14 at Cape Canaveral.**

Image 13

Scott Carpenter practices the lost at sea procedures that he actually needed.

Image 14

Wally Schirra leaving Hangar "S" on his way to LC-14.

Image 15

Schirra brought SIGMA 7 down so close to the carrier the splashdown was photographed.

Image 16

Gordon Cooper, at home in FAITH 7.

Image 17

**Rescue swimmers attending FAITH 7.
Cooper is still inside waiting.**

Image 18

and found fish flipping and swimming around in the bathtub! Not only had the perch thawed out, but they had come back to life! To a little boy of my age, that was far more interesting than that silver rocket on the TV. Still, I did take the time out to watch the launch, then went right back to fish-watching.

As the countdown drew into its final seconds, Mercury astronaut Scott Carpenter could be heard on the voice loop saying "Godspeed John Glenn." For Carpenter it was a brief, vocalized prayer for the well-being of his friend and fellow astronaut. If such were to occur today, however, sadly the person who spoke the words on a "government" radio would likely end up being sued by an atheist.

Finally just 39 seconds after 9:47 a.m. the count reached zero.

(See Image 11, page 84.)

The ignition sequence for the Atlas-D booster involved the light off of the two vernier steering thrusters on the sides of the vehicle followed less than one second later by the ignition of the three main engines. Although official documents state that liftoff occurred "… about T+4 seconds after ignition," examination of historic video appears to show liftoff initiated around 6.7 seconds after ignition. The precise moment of liftoff was probably not a large concern aboard FRIENDSHIP 7 that morning. What John Glenn was focused on was following to-the-letter the procedures that he and his fellow astronauts had written for the launch phase of the Atlas booster.

(OTHER Memories) Donny; the Space Coast, Florida: "Everyone in our school was outside listening to the countdown on some small radios. We heard about the liftoff and then looked over in the direction of Cape Canaveral. After what seemed like a long time we saw it! "There it goes" and "There goes John Glenn!" the teachers were shouting and pointing. The rocket itself was just a white sliver with a long tail of bright yellow fire. We watched it for a long time until it faded from sight. When I went home for lunch it was on the TV and they were showing where he was on a map. My mom was complaining that there were so many people in Cocoa Beach that you couldn't even drive and that the grocery store was completely sold out of bread and milk."

FRIENDSHIP 7:
SEGMENT 51

Glenn was well aware of the concerns that many engineers had about the Atlas booster itself. One such concern was his passing through the area of maximum aerodynamic pressure on the vehicle, or "Max-Q." It was the area of boosted flight where the previous Mercury Atlas, MA-2, had failed. Glenn's dilemma was that Max-Q and the sound barrier are normally relatively close to one another along the flight profile. The transonic region just happened to be an area where the maximum cabin noise and vibration took place on the Mercury Atlas. Dutifully, John Glenn reported that vibration by saying,

"… Have some vibration area coming up now." Then 28 seconds later he considered what he had just said and decided to reassure those on the ground, some of whom had their finger on the command destruct button.

"We're smoothing out some now, getting out of the vibration area," he reported.

At 00:02:09 into powered flight the two outboard booster engines shut down and were jettisoned. This point in the Atlas' launch was known as "BECO," or Booster Engine Cutoff. The remaining engine, known as the sustainer engine, continued to burn for the remainder of powered flight. John Glenn then saw some of the plume backflow from the BECO and mistook it as being his escape tower jettisoning. Some 25 seconds after the booster engines separated he witnessed the actual automatic jettison of his escape tower. Overall powered flight lasted for five minutes and one second.

CAPCOM Alan Shepard called FRIENDSHIP 7 and announced,

"You are go for at least seven orbits!"

Although he had been cleared for "… at least seven orbits," John Glenn's flight duration, according to the official NASA "News Release" of January 21, 1962 was planned for "… a one, two or three-orbit mission." Although many accounts of Glenn's return after three orbits rather than seven imply that the mission was cut short due to a problem with a heat shield indication, the fact is that the mission was planned for just three orbits. The extra four orbits were simply a contingency in case the spacecraft could not be brought down on orbit number three due to some unexpected problem. CAPCOM's term "at least" was likely just Shepard misspeaking the clearance which should have been stated as, "a maximum of seven orbits," which was actually what was in the flight plan.

Every ear in the free world and countless ears behind the Iron Curtain were glued to their radios as FRIENDSHIP 7 orbited the Earth. They listened intently as Shorty Powers described the flight and as radio reporters reinterpreted his words. There was the universal feeling being that "we" finally have a man up there. Now the distinction became obvious that the difference between the United States flying an astronaut and the Soviet Union flying a cosmonaut was that NASA conducted its flights under the glaring scrutiny of the news media and the full view of the public, while the Soviets conducted their flights under the shroud of totalitarian secrecy.

(OTHER Memories) Paul; the Midwest: "I remember my father reading aloud an article from the newspaper after Glenn's mission. It told about a judge in a courtroom who called a recess in a burglary trial so everyone in the courtroom could watch the flight on TV. The TV that they used was the one that the guy on trial had stolen- it was there as evidence. The judge ordered it plugged in and everyone watched John Glenn's flight right there in the courtroom! Father laughed so hard when he read that."

By the beginning of his second orbit Glenn was dealing with a persistent problem with a yaw thruster that was becoming something of an annoyance. A potentially more dangerous problem, however, was being indicated by the telemetry system. The "segment 51" sensor was indicating that the capsule's landing bag had deployed. Folded accordion-style between the heat shield and the capsule's aft bulkhead, the landing bag was designed to pop out after reentry and act as both a cushion and the sea anchor at splashdown. If the indication was valid, John Glenn's capsule would burn up on reentry. A firestorm of phone calls and loop communications discussing this issue took place following its initial reception by the Bermuda "TM" controller. Glenn had not a hint of the problem until the 02:00:47 point in the mission.

It was then that the Canton Island CAPCOM simply said "We also have no indication that your landing bag might be deployed. Over."

To which John Glenn replied "Roger. Did someone report landing bag could be down? Over."

The CAPCOM then blew it off by replying with "Negative we had a request to monitor this and to ask if you've heard any flapping, when you had high capsule rates…"

John Glenn then went on with his very busy work schedule paying little mind to the landing bag reference. Meanwhile at Mercury Control the serious debate was being waged between those controllers who felt that the Segment 51 signal was erroneous and should be ignored and those in NASA management who thought the best plan of action would be not to jettison the retro rocket package after retro-fire. The retro rocket package was held on to the heat shield and the spacecraft by three straps. Management felt that those straps may hold on just long enough to keep the heat shield in place.

There are a lot of accounts written by different individuals who were there at that moment in history when the flight of FRIENDSHIP 7 appeared to be in jeopardy. There are a lot of documentaries that have been created about that moment in history as well. Yet in most of those no one specifically says exactly who it was that came up with the idea to leave the retro package attached. In the largely fictitious movie "The Right Stuff." a German rocket scientist is shown holding a Mercury capsule and discussing the options. Just for the sake of reality and considering how steeped most of us are in pop culture, it is important to note that aside from a few launch pad operations there largely were no German rocket scientists involved with the

Mercury capsule itself. In fact, the one individual leading the charge for the retro package solution was the developer of the Mercury capsule; Max Faget. An original member of the Space Task Group, he carried tremendous weight in the Mercury Program. According to Gene Kranz's terrific book "Failure Is Not an Option," reentry with the retro package remaining attached to Glenn's capsule was Faget's idea. Considering Faget's influence as being the father of the Mercury Capsule, it is fair to conclude that Kranz's version of the story is the closest to being factual.

On the third orbit FRIENDSHIP 7 came into ground contact with the Hawaii station at the 04:21:00 mark of the mission. Just one minute and 45 seconds later the CAPCOM informed John Glenn of the full nature of their concern,

"FRIENDSHIP 7, we have been reading an indication on the ground of Segment 51, which is landing bag deployed. We suspect this is an erroneous signal. However, Cape would like you to check this by putting the landing bag switch in auto position, and see if you get a light. Do you concur with this? Over,"

Glenn replied, "Okay. If that's what they recommend, will go ahead and try it. Are you ready for that now?"

From that point on John Glenn was actually brought into the loop concerning the problem the ground controllers had been studying for more than an hour and a half. Dutifully Glenn moved the switch into the auto position.

"Negative, in automatic position," He then reported, "did not get a light and I'm back in the off position now. Over."

No one really knew for certain whether FRIEND-SHIP 7 would survive reentry. And although contemporary documentaries and movies may depict an agonizing length of time that John Glenn had to contemplate the problem, in fact, just 20 minutes and five seconds separated his notification by ground controllers and reentry blackout. At 04:42:50 into the flight FRIENDSHIP 7 began entry interface. Against the judgment of flight controllers, NASA management had elected to use an untested procedure based only on the Segment 51 indication. They ordered the retro package to be left on FRIENDSHIP 7's heat shield. It was their hope that the landing bag would remain attached through the majority of the four minute reentry. In fact, from Glenn's real-time reports, the retro pack remained attached for just the first 24 seconds of the reentry.

Following the reentry FRIENDSHIP 7 simply dropped through the atmosphere to a little more than 70,000 feet. It was kept stable during that time by the constant firing of the reaction control system jets. Finally, at an altitude of 30,000 feet, the drogue chute automatically deployed. This small parachute was designed to stabilize the capsule as it descended into the thicker atmosphere. The final step in the parachute sequence came passing through 11,000 feet with the release of the main parachute.

"Chute is out, in reefed condition at 10,800 feet," Glenn reported and then exclaimed gleefully, "and beautiful chute!"

Exactly 12 seconds after deployment of the main parachute the landing bag that had caused so much worry prior to reentry deployed normally. For the next five minutes and 10 seconds FRIENDSHIP 7 floated gracefully to the surface of the Atlantic Ocean where it splashed down and was safely recovered later by the Navy destroyer USS NOA.

FRIENDSHIP 7:
GLENN DID NOT NEED
ANY FLOTATION DEVICES

This time as John Glenn's spacecraft bobbed in the Atlantic, unlike Gus Grissom's LIBERTY BELL 7 capsule, FRIENDSHIP 7 had a flotation collar. In retrospect Glenn did not need any flotation devices- after that flight he could walk on water.

Euphoria across the free world immediately followed John Glenn's FRIENDSHIP 7 flight. Americans had the real feeling that they were finally gaining momentum in the space race with the Soviet Union. Although the Soviets were actually well ahead in flight duration and in lifting capability, NASA sought to trump that advantage with the fact that the United States launched their rockets in the full sight of the public while Russian flights were conducted in secret. Of course, John Glenn's flight was censored to a degree. The landing bag issue, which could have ended Glenn's life, was not disclosed until after the flight. By then, of course, euphoria of a successful mission completely overshadowed that little problem. Thus today, when you see depictions of Glenn's FRIENDSHIP 7 reentry where news people and the public are agonizing over whether or not Glenn would survive, that is completely incorrect. No one outside of Mercury Control and NASA's upper management had any clue that there might have been a problem as that reentry took place. It ended up that the fix for the landing bag was quite simple. The heat shield deployment limit switches, which indicated that the heat shield was released, were rewired in series rather than parallel and rigged farther away from the actuation points. The landing bag anomaly never reappeared on future Mercury flights.

More than a half-century after the flight of FRIENDSHIP 7 we remember and still celebrate John Glenn's successful mission. The FRIENDSHIP 7 flight moved people, it moved a nation and it moved the free world. It silenced congressional critics of NASA and it gave President John F. Kennedy a shining success in his effort to reach the moon. Immediately after the flight every kid in the United States knew who John Glenn was, what he had done and what Project Mercury was. It had been demonstrated that the United States could indeed put a man into orbit and bring him back alive. We had the spacecraft, we had the booster, we had the infrastructure, we had the manpower, we had the NASA leadership and we had the national will to do the job.

Following his flight, John Glenn found himself somewhat soft-core grounded from future space flights. NASA and the White House knew that they had an American icon on their hands and were unwilling to risk him on another flight. He was being prompted repeatedly to take a management position within NASA. Still he kept training with the new class of astronauts and conducting ground support with his Project Mercury brothers in the "Original Seven" group of astronauts. Eventually, he decided to leave NASA and seek outside career ventures. He did appear with Walter Cronkite on a CBS television special called "T-Minus Four Years, Nine Months and 30 Days." At length, John Glenn went on to serve his nation as a United States Senator from the state of Ohio and while doing so managed to gain a seat on the Space Shuttle. In 1998 he became the oldest human to ever fly in space.

AURORA 7:
IDIOPATHIC ATRIAL FIBRILLATION
AND ITS RESULTS

The next flight up was scheduled to be flown by Deke Slayton and would be a mirror to Glenn's mission. Of course, within the general public, no one really knew much about that upcoming mission. All that anyone was "talking about" was John Glenn and FRIENDSHIP 7. Even a five-year-old like me knew the name John Glenn. However, what no one outside of NASA knew or expected was happening. One of the seven Mercury astronauts was about to get quickly swept under the political rug.

On March 15, 1962, Deke Slayton was removed from flight status due to a minor heart issue called idiopathic atrial fibrillation. A brief press conference was held, a few news stories were published and broadcast and that was that; Deke Slayton was out as far as the public was concerned. But, we still had John Glenn, Alan Shepard and that other guy whose capsule sunk, so what the heck.

NASA had been aware of this minor fluctuation in Slayton's heartbeat since the beginning of his days with the agency, so why had it come up now? A rumor that John Glenn had a heart issue began circulating in high places around Washington, D.C., about a week prior to his flight. When NASA squashed that rumor, another popped up claiming that one of the Mercury astronauts had a heart problem. When NASA Administrator James Webb was handed the rumor plus the information that it was Slayton who had the "problem" he ordered an investigation into Slayton's medical records. Weeks later a small army of doctors descended upon Slayton and unanimously concluded that, although he had this minor skip in the rhythm of his heart-beat, it was not a real problem. Webb was not satisfied and so a three-doctor panel was assembled. After giving Slayton a very superficial examination they met with Hugh Dryden, who then gave Deke the bad news; he was off the flight. James Webb had a Mercury 7 astronaut's scalp on his belt, which got some powerful persons in Washington off his back.

Slayton was offered and accepted the job of Coordinator of Astronaut Activities in July of 1962, but the formal announcement was not made by NASA until a few months later. Slayton went on to become one of the most influential people in the history of the US manned space program. Finally, in July of 1975 he got the chance to fly in space aboard the Apollo Soyuz Test Project. After that he remained a powerful force within NASA. Although he had been shafted by just a couple of doctors, without his keen management skills and people sense as well as pilot judgment America's space effort would have probably suffered to a great degree. In spite of their myopic intentions that hand full of doctors did the program a huge favor.

(See Image 12, page 85.)

When Slayton was removed from his flight, which would have been named "DELTA 7," Walt Williams, who was the Flight Operations Director for Project Mercury, was forced to select a replacement. Wally Schirra had been assigned as Slayton's back-up, but Scott Carpenter had just finished months of training as John Glenn's back-up and Williams took that into consideration. He also knew that an eight-orbit mission was coming up after Slayton's vacated mission and he felt that such a mission would be perfect for Schirra. With that in mind, Scott Carpenter was selected to fly the second orbital Mercury mission.

So it happened that three months after John Glenn's three-orbit flight, astronaut Scott Carpenter was prepared to repeat the same mission. The project Mercury spacecraft that would take him into space had been christened "AURORA 7." Considering the everlasting fame and notoriety that Glenn's flight generated, Scott Carpenter's flight would be known throughout the years mostly for one thing in particular: the fact that his spacecraft overshot the landing point by 250 miles. A few of the people involved in the AURORA 7 mission have, in their autobiographical accounts of the era, attempted to blame Carpenter for the landing overshoot and a number of other issues that cropped up during the flight. At least one person has gone as far as to imply that Carpenter was incompetent as an astronaut. When you read the following account of his mission, however, you will likely see things much differently. It was management that was incompetent, and not the astronaut.

In the original Project Mercury schedule, Glenn's flight, MA-6, and the next flight, MA-7, were to be identical missions lasting three orbits of the Earth. At that time NASA was being extremely careful with its orbital test flight program and its astronauts. The ideal was to get up, fly the vehicle and get back down without any extraneous tasks. As a matter of fact, John Glenn often states that on his mission, "…they didn't even want me to bring a camera. They were that afraid I'd get distracted." So adamant was NASA management to not give Glenn a camera that he had to go out and buy one himself. In the wake of the success of Glenn's three-orbit flight, however, there came a relaxation in that way of thinking and it opened the doors to some new areas of thinking. It was felt that now that we were able to fly into space it was time to add a large degree of scientific study onto future flights. All of this started to be dumped upon the original pilot assigned to MA-7… Deke Slayton. In his autobiography "Deke!," he wrote about the push to inject scientific experiments into the flight plan saying,

"…everybody and his brother came out of the woodwork… One damn thing after another. I had my hands full trying to resist it."

Fortunately for Slayton, when he was removed from flight status about a month after Glenn's flight, this problem went away. It would be up to the next pilot to handle the new "science" workload and that pilot turned out to be Scott Carpenter.

Almost no one in NASA management bothered to consider the fact that Mercury had only flown once into orbit as a manned spacecraft. This was a new vehicle operating in a new place for humans, it was flight test on the highest order and no place to be loaded down with all sorts of experiments. However, unlike Slayton, Scott Carpenter was enthusiastic about adding all of the "science." As it turned out, this mixture of a new program, a new spacecraft and a three-orbit flight peppered with experiments was an unhealthy combination.

At length, the experiments added into the MA-7 flight included a multicolored balloon released from the spacecraft for observation and photographic purposes, a ground flare visibility experiment, observations of the air glow layer of the atmosphere, a "zero gravity experiment" consisting of three glass balls with a small amount of liquid in them, a laundry list of photographic experiments and assorted star sightings and measurements of star brightness, as well as personal observations of auto kinesis and spacecraft reaction to movement of arms and legs within the cabin. What's more, while Carpenter had spent 70 hours and 45 minutes in the simulator training for his specific flight, almost none of that time was spent simulating in detail working with the experiments and conducting spacecraft operations at the same time.

It just so happened that around this same period I got my first big break into the news media spotlight and the world of hard core journalism. As the school year was winding down in the city of Saginaw, Michigan, enrollment in kindergarten for the next fall began. My mom walked me some 80 feet from our home on Parkwood Street, and across the expanse of Webber Street, to the huge brick building were Webber elementary school and junior high school were both housed. Once inside, she worked at shuffling all of the paperwork, while I scribbled on some spare sheet of scratch paper to keep myself occupied. While we were there a reporter from the Saginaw News dropped in and came to the table to shoot a few photos. The following day the paper arrived and there was my picture! A few days later our neighbor, who was a news photographer, gave my mom the four-by-five inch glass plate that was used to print the photo onto the papers; I still have it today. Indeed there was great excitement. At the grand old age of five I was in the news! Of course, just a few days later, Scott Carpenter would completely overshadow me by flying into orbit. Gee, we Americans have short attention spans and are so easily distracted. No wonder fame is fleeting, especially for a five-year-old.

AURORA 7:
OVERLOADED

At 07:45:16 EST on the 24th day of May 1962, the MA-7 flight lifted off from Cape Canaveral. The Atlas booster performed perfectly and AURORA 7 was inserted into the correct orbit.

(See Image 13, page 86.)

Then at 00:08:27 into the mission Carpenter reported that he was "...beginning to unstow the equipment." Shortly after that the astronaut fulfilled the first of his non-flying tasks by taking a series of photographs of the expended Atlas booster. By the 00:16:19 point of the mission the first hints appeared of a problem that would plague Carpenter throughout the mission.

"I think my attitude is not in agreement with the instruments." Carpenter reported.

What he was seeing was a beginning of a degradation in his Automatic Stabilization and Control System (ASCS). Just three minutes later the first signs of the pilot being overloaded began to show.

While in contact with the Canary Islands station and receiving a read-up of times for sunset and sunrise, Carpenter was too busy working with his camera and was unable to copy the numbers. Although he had practiced that moment of flight many times in the simulator, he had never been required to do the task in zero gravity while wearing a pressure suit and gloves and manipulating a camera that had never been designed to be manipulated under those conditions. Further evidence of him beginning to be overloaded came at the next station passage at Kano, Nigeria.

"I am a little behind in the flight plan at this moment. I have been unable at this time to install the MIT film. I finally have it. I'll go through the gyro caging procedures very shortly." Carpenter replied when asked if he was doing his gyro caging procedures.

That was at 00:26:37 into the mission, and Carpenter was already falling behind. None of this was actually Carpenter's fault. After the flight he reported that the procedure for caging the gyros, although practiced in the simulator, was not done with live equipment and a visual presentation. He found that the procedure, which he had often done, now ate up valuable time. He had similar difficulty loading the film in and out of the camera with the spacesuit and gloves on. Likewise he found there was nowhere near enough time to allow for the transition from one activity or experiment to another. Ultimately Carpenter was forced to stow the camera, forgo the horizon pictures and focus his efforts on the gyro caging procedures. It is important to note here that the seven Mercury astronauts had spent years training in any way that they could to try and prepare for flying in space. Everything from flying aircraft in parabolic arcs to swimming in SCUBA gear and navigating by compass in the Chesapeake Bay went into their training. But no one really knew what tasks in space would be easy and which would be hard. So, Carpenter's difficulty was caused by something that NASA had simply overlooked in the training of the astronauts. The training for these operations could now be changed. Such was the real purpose of project Mercury.

Later on in this first orbit Carpenter was unable to perform an observation to measure the oscillation of the light from Venus as it came through the horizon. The simple act of removing the equipment and getting it set to go while wearing gloves took so long that by the time he had the equipment set Venus was well above the horizon. Then, as Carpenter focused on the Venus experiment, he inadvertently left his automatic and manual thrusters on. In the Mercury capsule this "double authority" consumed fuel at a huge rate. Fortunately, Carpenter caught the error relatively quickly, but it still cost him far too much Reaction Control System "RCS" fuel. During his first orbit Carpenter reported three times that he was behind schedule for flight activities.

Early into the second orbit over the Canary Islands station Carpenter was told that Mercury Control was worried about his auto fuel and manual fuel consumption. He reported back that he was concerned too and he would try and conserve fuel. As if to add to the workload, telemetry concerning the temperature of Carpenter's suit continued to give ground controllers false indications. Thus, at nearly every reporting station the controllers would read up their numbers which Carpenter would debate. At one point they showed a suit temperature of 102 degrees Fahrenheit, while Carpenter himself sat comfortably in about 74 degrees Fahrenheit. Over the Indian Ocean Carpenter figured out that his ASCS system, which he had been relying upon to conserve fuel, was probably acting up and may be costing him additional precious fuel.

Shortly after passing out of the range of the Indian Ocean tracking ship, Carpenter deployed a small tethered "balloon" from his spacecraft and then did some recording on his onboard tape recorder. He was making notes as to his current laundry list of issues – from his suit temperature to his doubts about the ability of the AURORA 7 spacecraft to hold its stability.

"I have gotten badly behind on the flight plan now," he also stated quite flatly.

When Carpenter came into range of the Muchea, Australia tracking station his CAPCOM was Deke Slayton who passed on an order as well as a stern warning from Mercury Control.

"We suggest you go to manual at this point and preserve your auto fuel." Slayton directed.

Then, following some standard banter about the astronaut's general condition, Slayton warned Carpenter.

"For your information, Cape informs that if we don't stay on manual for quite a spell here, (we) will probably have to end (on) this orbit."

What this was in effect saying was that Carpenter's decision to remain on auto was incorrect. The ASCS was clearly malfunctioning and in doing so draining AURORA 7's fuel. Fortunately the mission had just moved into a phase where drifting flight was required. Carpenter switched to manual, took his hands off the controls and begin to drift in what Wally Schirra later called, "chimp mode."

For almost the complete remainder of his flight Scott Carpenter simply drifted along the orbital path in which he had been inserted. In looking at the post-flight fuel consumption charts it is clear that almost any maneuvering at this point would have resulted in Carpenter running out of fuel. Aboard AURORA 7 the auto fuel warning light had been on for such a long time that Carpenter actually covered it with a piece of tape to keep it from annoying him.

Passing over Hawaii something cropped up that drew Scott Carpenter's attention, perhaps more than it should have. Back during John Glenn's historic flight Glenn had discovered something he called "fireflies" swirling around his spacecraft. The single observation ignited a debate in the science community as to exactly what these little white objects, officially named "space particles," might be. Some people went so far as to speculate that these so-called fireflies may actually be living creatures. So far during the mission of AURORA 7, Carpenter had seen only a very few individual particles that seem to drift by. Now, as the astronaut moved in the cabin to attempt to cage and test his gyros, he suddenly saw a great number of the "fireflies" outside his window.

"They definitely look like snowflakes this time." Carpenter reported. He then added that he would attempt to photograph the particles.

Given a "Go" for his third and final orbit, Carpenter's next task was to jettison the "balloon" that he had deployed earlier in his flight. The balloon experiment, which was to investigate visual phenomena in the space environment, simply didn't work out. Using a 30 inch Mylar sphere that was to be inflated to 900 psi, the experiment required the astronaut to observe panels on the balloon painted orange, white, silver, yellow and phosphorescent. The balloon itself was attached to a 100 foot nylon line. On the other end it was attached to the spacecraft with a "strain gauge." It was hoped that measurements of the balloon's tension on the capsule could be measured. Additionally it was hoped that the balloon, as it unfolded, would dispense a quantity of 1/4 inch Mylar discs or "confetti" that had been placed into the folds when the balloon was packed.

Officially the description of the experiment was: "The astronaut will observe the operation from the deployment sequence, through tethering, to release and separation, and any oscillations or gyrations will be noted. Photography of angular displacement, the various colors, and the confetti dispersion will be provided for correlation with visual responses. The astronaut will orient the spacecraft in order to track the balloon's trajectory after it is released and photographs during this phase are requested when distances are recorded."

Scott Carpenter did not have the camera dexterity or the fuel required to do any of that. The confetti was never observed, the balloon did not fully inflate and the tether holding the balloon to spacecraft rarely had any tension. Normally the balloon simply floated randomly around out of sight of the spacecraft window and the tether was often so loose that it looped. Carpenter made only one observation of visual acuity saying that orange was the color most easily seen. Then, at the beginning of the third orbit over Cape Canaveral, when he went through the procedure to jettison the balloon it failed to let go. It remained attached until reentry when it simply, "went away."

Cape Canaveral CAPCOM asked Carpenter to give them an observation of the zero-G experiment as he headed out over the Atlantic on the final orbit. The astronaut dutifully gave his observations which CAPCOM was unable to clearly receive, so it had to be repeated shortly before AURORA 7 passed out of range. Switching to the capsule's voice record mode, Carpenter went on to describe zero-G, how it affected his eyes and the lack of auto kinesis. Then Carpenter used some of the time on his last pass over the Atlantic Ocean to unstow and again load the MIT film into his camera. This time he was determined to get the photographs that he had missed on his first orbit. He also took time to talk more about the fireflies and his own spatial orientation. Oddly, there he was on the brink of running out of fuel that would be critical to maintain his proper orientation during reentry, yet Scott Carpenter's greatest concern was cramming to get all of the "science" done. It was very clear that Scott Carpenter had fully bought into the added science workload and was determined to get as much done as he could.

An additional load continually was placed upon Scott Carpenter by his highly dysfunctional spacesuit cooling system, which seemed to require constant adjustment and readjustment, all of which had to be reported to ground controllers. Additionally, the flight surgeons on the ground required a blood pressure measurement at nearly every station. Carpenter's own confidence and attitude, however, remained quite good. When informed by ground controllers that his respiration data, which had previously not been received on the ground, was finally being received he said,

"That's very good. I guarantee I'm breathing."

During the remainder of this third and final orbit Carpenter managed to get a respectable portion of the remaining science experiments checked off his list. Then at 04:19:22, sunrise occurred and he found himself in the same luminous swarm of fireflies that John Glenn had witnessed. He also noticed that if he rapped his hand on the hatch he could stir up thousands of them. Scott Carpenter had solved spaceflight's newest mystery: the space particles were nothing more than ice crystals clinging to the capsule.

Approaching Hawaii for the final time, the absolute most critical portion of AURORA 7's flight was about to begin: retrofire. Carpenter was instructed to orient the spacecraft and go to ASCS in preparation for the pre-retro sequence. He accomplished that task, but then diverted his attention to the stowage checklist.

A moment later he reported to the ground,

"… my control mode is automatic and my attitude standby, waiting minute, I have a problem in…"

Ground controllers waited 33.5 seconds before Carpenter continued,

"I have an ASCS problem here. I think ASCS is not operating properly."

Moments later he returned to fly-by-wire controls and reported that his attitudes did not agree. He was five minutes from retrofire. He also reported that he was showing a descent rate of about 12 feet per second. Hawaii ground had just enough time to run through the pre-retro checklist with him before he went out of range.

"I don't have agreement with ASCS in the window Al. I think I'm going to have to go to fly-by-wire and use the window and the scope." Carpenter reported to CAPCOM Alan Shepard upon contacting the California station.

Shepard agreed and told Carpenter he had 30 seconds until retrofire.

Sources tend to disagree on the exact error in yaw that was present during the retro sequence and reentry, but the best figure seems to be 25 degrees. This overall error was a combination of gyro error, ASCS error and the inability of the human eye to precisely align yaw using the Mercury spacecraft window and its periscope. This error, however, was responsible for a large portion of the landing overshoot. Another portion of the overshoot was caused in the mechanical process of triggering the retro rockets. Carpenter reported in his post-flight interview that when the retro countdown reached zero he thought he did not get ignition, so he manually punched all three retro buttons in sequence. The time that it took between his evaluation that he had not gotten retrofire and his ability to react and push the buttons, in addition to the moments that it took to ignite the retro rockets, all added up to about 3 seconds. That equated to an overshoot of about 20 miles.

Carpenter also has commented over the years that he did not feel that the retro rockets were burning with the strength that John Glenn had described and, in his opinion, that also had added to the overshoot. During the actual retrofire Carpenter did report to the ground that there were issues concerning the burn.

"Okay, I think they held well Al," he radioed, "I think they were good. I can't tell you what was wrong about them because the gyros were not quite right."

It is however worth noting that none of the published historical records indicate any anomaly in the strength of the retro rockets' burn.

Nearly 10 minutes before AURORA 7 touched the Earth's atmosphere Scott Carpenter ran out of manual RCS control fuel. He now had no choice other than to rely upon his malfunctioning ASCS system during reentry. Just prior to entry interface he reported the fuel was,

"… 15 auto, I'm indicating seven manual but it is empty and ineffective."

During the reentry blackout Carpenter reported into his onboard recorder,

"I've got the orange glow. I assume we are in blackout now."

He put out a blind call to Gus Grissom at the Cape but got no return.

"There goes something tearing away," He continued his commentary saying, "Okay. I'm setting in a roll rate at this time. Going to AUX damp. I hope we have enough fuel. I get the orange glow at this time. Picking up just a little acceleration now."

Throughout the entire reentry phase Carpenter's ASCS kept him alive with a lot of help from Max Faget. Max, as the primary designer of the Mercury capsule, with great sense of forethought, designed the capsule with its center of gravity offset so that at any time in freefall within the atmosphere the capsule would always orient itself blunt end forward. Thus, even when AURORA 7's fuel tanks were empty and its thrusters inert it still dropped through the atmosphere in a relatively stable manner. There was, however, a fair degree of oscillation in the upper atmosphere and in order to dampen that out Carpenter wisely elected to manually deploy the drogue parachute. At the 10,000 foot altitude, which was preset in the spacecraft's barostats, the main parachute deployed, reefed and then fully opened to lower AURORA 7 to a safe splashdown.

On his way down to the splashdown point some 250 miles off target, Carpenter made broken contact with Gus Grissom at Cape Canaveral. There was just enough time for both to understand that his condition was good, his descent normal, and he was way off target.

AURORA 7:
MUCH ABOUT SHORTCOMINGS

Following splashdown Carpenter did exactly what he was trained to do. He exited the spacecraft through the upper bulkhead, tossed his life-raft into the water, jumped into the ocean, climbed into the raft and waited for the rescue force to come and pick him up.

(See Image 14, page 87.)

What followed was what Walter Cronkite once described as,

"perhaps NASA's greatest blunder in the history of manned spaceflight."

From the beginning of blackout until Carpenter himself was recovered nearly two hours elapsed and NASA told the news media almost nothing other than the AURORA 7 would likely land much farther downrange than expected. As far as anyone in the news media knew, America had lost Scott Carpenter.

Of course Scott Carpenter was recovered in terrific condition. He had spent the time sitting in his little yellow life-raft toying with a small fish that was his only company.

"I had sort of a blessing there for the hour after the flight," Carpenter later recalled, "Everybody else had been confronted immediately with a debriefing team, and that's an occupational hazard. So, I climbed out,

I got in the life raft, and I had a quiet time to contemplate what happened, and I treasure the recollection of that. Pretty soon- and I wasn't worried, either, because there's a SARAH (Search And Rescue And Homing) beacon that sending out signals to a lot of people listening, and I just didn't even think about it. But pretty soon the plane turned up. It was a plane I used to fly, and I waved to it. Then another plane turned up, and there were, before I was picked up, I think, seven airplanes flying around me, and I got tired of waving at them. I was sitting there in the raft and I heard this calm voice say, "Hi, there." And three Navy SEALs had jumped out of one of the airplanes and swam up to me. They had a big raft to put around the spacecraft. So we talked a little bit and I offered them some of my survival food. They said they weren't hungry."

The news media soon got over NASA's slight and Americans were just happy to have another space-flight hero.

Summing up the flight, it is pretty easy to see that the primary difficulty with the voyage of AURORA 7 was that NASA sent aloft a fairly new type of vehicle, on only its second manned orbital mission, with a pilot who had been heavily overloaded with tasks other than flying, and who had also been ill trained to perform those tasks. Scott Carpenter was overloaded from the first moments of his mission. Thus, who is to blame for any shortcomings in the mission? Chris Kraft in his book "Flight" heavily contends that Carpenter himself was incompetent. In examining the actual flight through historical records, rather than self-serving personal whims, it becomes clear that Carpenter was indeed competent and that the incompetent persons were actually in NASA management. This can be said because it was NASA management who allowed, what should have been a three-orbit test flight, to be infested in the last days prior to launch by a wide array of "science experiments." NASA management should rightly have frozen the flight plan and not allowed these extraneous tasks to be injected into the mission. Additionally, the Mercury spacecraft itself had some weaknesses and some defects that cropped up during the mission. Those alone were enough to keep an astronaut busy for three orbits. It is said that Carpenter got distracted by the "fireflies" but those were only a momentary excursion. It was the load of "science" that became the primary distraction. This was America's second step into orbit and it was far too soon to allow the mission to become a science project.

With hindsight we can look back at those first steps in human spaceflight with a sense of wonder at the courage that it took to climb aboard those tiny spacecraft. To those of us growing up with spaceflight the Mercury Seven astronauts were held up as heroes and elevated to a rightful position where they remain to this day. Scott Carpenter did a good job and his flight taught us as much by its shortcomings as by its successes.

SIGMA 7:
ARE YOU A TURTLE TODAY?

Often in spaceflight documentaries it is portrayed that Wally Schirra's Project Mercury flight, SIGMA 7, took place during the Cuban Missile Crisis of October 1962. Statements such as "Wally blasted off into the heart of the Cuban Missile Crisis" or "unlike the other Mercury flights, Wally splashed down in the Pacific otherwise he'd have landed smack in the middle of the Cuban Missile Crisis" and "No one hardly noticed his flight because the American public was riveted to the Cuban Missile Crisis" are often scripted into modern accounts of Project Mercury. This is proof that the production companies who make such TV documentaries have lots of money for hiring producers, assistant producers, assistant to assistant producers, but never spend a dime on research historians to fact-check their scripts. The fact is, the American public, the Kennedy Administration, the Department of Defense and even the CIA were unaware of a "Cuban Missile Crisis" at the time of Schirra's flight because there was no Cuban Missile Crisis at the time. SIGMA 7 flew on October 3, 1962, and the events of the Cuban Missile Crisis would not even begin for another two weeks!

SIGMA 7 was the first Mercury flight to be extended beyond Project Mercury's original design planning of just three orbits. This brought into play several problems, the greatest of which was the Earth rotating on its axis. During a three-orbit flight, such as that of John Glenn and Scott Carpenter, the earth rotated about 67.5 degrees underneath the orbital plane into which the spacecraft had been inserted. Thus, NASA had placed all of its ground tracking and communication assets within range for such flights. On Schirra's six-orbit flight, however, the Earth would rotate a total of 135 degrees under SIGMA 7. That meant that Schirra's spacecraft would work its way farther and farther south and would be out of range of more and more of the established stations on orbits four, five and six. Additionally, a retro-fire in orbit six would splash SIGMA 7 down in the Pacific rather than the Atlantic. So, Schirra's landing in the Pacific Ocean had nothing to do with the Soviet Union or Cuba; it had everything to do with the Earth's rotation.

Another problem that had to be solved was the fuel consumption issue that had bitten Scott Carpenter, who ran out of reaction control fuel during his reentry. Part of the blame for Carpenter's woes can be attributed to human error, poor pre-flight management and work-load planning, but much of the blame can also be pinned on system design. A Mercury spacecraft had two reaction control system (RCS) settings. First there was the Automatic Spacecraft Control System, also known as "Fly-By-Wire" or the ASCS. Second there was the manual system. Acting proportionally to the astronaut's control handle movements, the RCS system fired two sets of thrusters; a one-pound thrust series of thrusters and a 24-pound thrust series. On vehicles previous to SIGMA 7 it was possible to not only fire RCS in the ASCS mode, but also the thrusters could fire in manual mode and ASCS simultaneously – thus the pilot could gain "double authority" on the RCS if needed. The problem was two-fold in that the double authority configuration could be inadvertently used, as happened with Carpenter's flight, and there was also no way to isolate the 24-pound, high-thrust jets, which were fuel guzzlers. In SIGMA 7 the astronaut was given a switch to isolate the 24-pound-thrust jets and given "extra training" to prevent from accidentally entering the double authority mode. For most of his flight Schirra used only the one-pound thrusters in an RCS firing method known as "ASCS Low," that he liked to call "mouse fart."

Atlas booster number 113D was scheduled to boost Wally Schirra on the 18th day of September 1962. That date, however, was pushed back because that particular booster had been modified with a baffled injector in its engines. That modification was intended to eliminate the usual two second hold down delay following ignition. The Air Force, however, had experienced several failures with that modification and did not want the 113D booster to go directly from the assembly line to launching the mission without a static firing. The static firing process delayed the SIGMA 7 launch date until September 25. Before the September launch date could be reached Atlas 113D sprang a fuel leak, the repair of which delayed the launch until the third day of October.

(See Image 15, page 88.)

On the morning of October 3, 1962, at 07:15:11 a.m., Mercury Atlas 113D lifted off from Cape Canaveral. The boost was about as normal as anyone could have wished – in fact there was even time for a few jokes.

At 00:03:39 into the boost, CAPCOM Deke Slayton asked Wally Schirra, "Are you a turtle today?"

It is a classic question from the 1950s and 1960s that if the person who is asked does not properly answer, they are required to buy the drinks. Schirra told Slayton that he was switching his radio from VOX "transmit" to the onboard "VOX" tape recorder.

"You bet your sweet ass I am," Schirra answered.

That was the correct answer. He then went back to "transmit" and informed Slayton.

"Just trying to catch you on that one," Deke quipped.

All of that funning took place as the Atlas was boosting above 300,000 feet at speeds greater than 12,000 feet per second. These guys were test pilots, no doubt.

SIGMA 7:
ONBOARD RECORDING

SIGMA 7 was lofted into a 176 by 100 statute mile orbit that took the spacecraft around the Earth once every 88.9 minutes. Although Schirra did have a few "scientific experiments" to do during his mission, he was not completely overloaded as had been the case with Scott Carpenter. Schirra was tasked instead with actually test flying the equipment. The first major portion of that task was working with his space suit's environmental control. In Carpenter's flight the suit controls and downlinked telemetry tended not to agree, while the suit temperature fluctuated to a great degree. Working the problem soon became a major distraction. In the Mercury program itself, the job of working suit development had been Schirra's, so now the right man for the job was in the cockpit. He found that simply by making very small adjustments and allowing the suit's system time to catch up there was no problem at all.

Schirra also made good use of his onboard tape recorder capturing much of what he saw and felt. On his first orbit, between Hawaii and Guaymas, Mexico, Schirra spotted the "fire flies" that both John Glenn and Scott Carpenter had observed. He narrated into his tape recorder:

"I'm starting to see the sunrise in the periscope. First light in the periscope during this particular orbit as a result of the night side. It is obvious that the periscope has no function whatsoever in retro-attitude on the night side. First light that I get is right now at a CET (Capsule Elapsed Time) of practically 1:25... The sunrise is coming in rather rapidly through the periscope. I do have the lighted objects that John mentioned, and I can create some by knocking them off. I definitely had a sensation of their being a field and of varying in size from small to bright. The periscope itself is blinding me. I'll have to put the chart on it, so I can see out the window. I am in condition for retro at any time, so I have nothing else to do but look out the window, assuming that the suit circuit is satisfactory. That chart helps no end to cover up that blasted periscope. Quite a large field of these objects. Definitely is confirmed that you can knock them off the hatch as Scotty said. And they stream off at, definitely there is no problem in judging that they are going away from the capsule, at a different rate than you are. They are definitely going slower, in velocity, than the capsule itself. One rap, and you can see them sliding aft. They are too small an object for photography. I would not even attempt to take a picture of them. Retro-attitude is being held very well by the ASCS."

Although Schirra's recorded dialogue may seem trivial to us today, in 1962 very little was known about man in space. Thus, even the smallest of comments by this third astronaut in orbit was considered to highly important. Scientists and engineers on the ground who were designing future spacecraft were keen to hear even the smallest sensation experienced by the

Mercury astronauts. Every word from Schirra was important and he had been trained to record every thought and sensation.

SIGMA 7's second orbit found Schirra working to see how well he could adjust the spacecraft's alignment and attitude without the use of his instruments, the primary concern being in the yaw axis. In the first test, Schirra covered his attitude indicator and went into the retro-fire attitude using features seen through the spacecraft window. He then yawed eight to ten degrees away and attempted to get back to the initial attitude. He later repeated the exercise by yawing 23 degrees out and returning. Each time he came to within two degrees of his starting point.

Keeping in mind that everything he experienced during his flight was important to the NASA engineers, Schirra continued to make good use of his onboard tape recorder. He had some free time to talk to the tape a bit more at 02:17:00 Mission Elapsed Time, after passing out of range of the Indian Ocean ship on this second orbit.

"Sunset is rather striking." Schirra recorded, "I don't think that I need to waste much time looking at them. They're very interesting. The other thing... it's fascinating how black it is when your eyes are not adapted. I definitely can see some coating on the window. Going back to VOX at this point in UHF-high to transmit. Opening visor to wipe off the right microphone. I licked it."

Later he went on to settle a bet,

"Cabin heat exchanger is evidently selling out at 40 degrees, and it's given me a hard time because I obviously owe Frank Samonski 50 cents."

On his third orbit Schirra was asked to try the "eating test" and he also went ahead on his own to conduct the "Voas test." This test was named for Dr. Robert Voas who had worked extensively in forming the aero-medical training for the Mercury astronauts. His test consisted simply of having the pilot close his eyes and reach out for known objects in the cockpit and thereby see if there was any disorientation in zero gravity. Many involved in the Mercury program objected loudly to this test. They were in fear that the astronaut may hit a switch or lever and accidently abort the mission – or worse. By the time of SIGMA 7, however, the timid managers had been quieted to the point where the test was allowed. Schirra reached out and with his eyes closed touched the manual lever, the clock, some rivets above the clock, the yaw indicator and his emergency rate lever without any disorientation and without any disaster. He spent the rest of that orbit flying attitudes and sighting stars.

As SIGMA 7 entered its fourth orbit, the Earth below was slipping farther north. NASA had pre-positioned tracking ships in the southern Pacific to alleviate the communication and tracking gaps. Stationed near Midway Island were the vessels HUNTSVILLE, AMERICAN MARINER and WATERTOWN. Additionally the Atlantic relay ship was refitted with command capability and stationed south of Japan and the Quito, Ecuador tracking station was activated.

Schirra spent orbit four powering systems on and off to see how they acted when exercised. He also took photos and practiced additional star sighting and vehicle orientation. His was a text book example of a test pilot taking her up, ringin' her out and bringin' her back – exactly the sort of mission Wally Schirra loved to fly.

On the fourth orbit between Cape Canaveral and the Indian Ocean Tracking Ship, Schirra dictated into his tape recorder,

"At this point in time," he recorded, "which is of course just prior to sunset, we're coming up with a batch of white particles. They show up in the blue sky. I have the horizon almost in sight. And they are drifting away from me. Let's check and see if we actually do get yaw out of these. They're tending to go up in relation to me, rather than tending to draw away. With this kind of lighting I can really see the illusion of visibility, due to the external problem of having smoke on the outer (window) panel. Definitely is not on the inner panels. It is quite easy to see by changing panels through reflections that it's the outer panel. We are now going into night coming up on five hours 11 minutes." He later added "Here we go into night rather rapidly. Now we're getting into the night side. I am apparently pointed towards the surface of the Earth, as I can see clouds with lightning in them."

SIGMA 7's pass between the Cape and the Indian Ocean Ship on orbit number five gave Schirra yet another chance to make a series of reports to his onboard tape recorder.

"… I'm slowly but surely coming into retro-attitude," he described as he manipulated the SIGMA 7 spacecraft, "All axes are working very well. Setting up roll. Getting your rates, pow, pow. And I want this to count… I'm going to go back to fly by wire now… That was stupid. Now we go to fly by wire low. I had a case of double authority and I really flotched (sic) it.

But better conserve our fuel. It's much too easy to get into double authority, even with the tremendous logic you have working on all the systems..."

It was fairly clear how easy it was in the Mercury spacecraft to slip into the thruster double authority that Scott Carpenter had experienced. Even a pilot who was not loaded down with egghead science could get into the trap.

On his final orbit, after passing out of range of the Quito, Ecuador station, Schirra took a bit of time to shoot some final photos.

"Very interesting terrain pictures." Schirra noted, "I'll take one of the horizon just for posterity. At this time that picture was A number 12, resetting to B and now have B-1. Taking some colored pictures of the South American continent. I don't think we'll have much luck with them."

Just 10 years after SIGMA 7's pilot took rudimentary pictures of his orbital views, astronauts would stand on the lunar surface and take photos of rocks.

SIGMA 7:
PROOF THAT MOUSE FARTS WORK

Firing SIGMA 7's retro rockets at exactly 08:52:21 Mission Elapse Time, Schirra had just under nine minutes before blackout began. At the time of blackout SIGMA 7 was passing nearly directly over the naval vessel USS WATERTOWN.

"I have selected aux damp and rate command at this time," Schirra said dutifully, as he began narrating the reentry events to his onboard tape recorder, "The window is almost completely occluded. It would be impossible to see out of it at this point. I'm seeing things come off, but I can't see them very clearly. There we go into the .05, a green. (The .05G light indicates the first tastes of the upper atmosphere as the spacecraft begins reentry) I am hands-off at this point. In rate command, in aux damp. And I have a roll rate started. A slight pitch rate, not bad at all. I can see out the window, for some strange reason, at last. There goes another long spiral-like looking device. I will give another blood pressure at this point, subsequent to .05G. All rates are very nominal. Rate command is working quite well I would say. Going back into G-field. And the attitude looks very stable. I'm rolling right around the horizon. I'm going to stop my blood pressure at this time... And sit back here and regroup. I can see the ion layer.

I'm inverted this time. Attitudes are controlling very well. Seems to be plenty of manual fuel... I'm still at 72 percent. Definitely has the cyclic rate in pitch at this point. Yaw is fairly stiff, G is building up. Capsule is quite stable. There is a green flow... and looks like orange streaks every once in a while. RSCS is doing very well on reentry. Rather unusual slow roll. Building up to G's. I have plenty of fuel in rate command. Seeing sparkles coming by now. A definite green glow, like limeade; G's building up. Oscillations are very good at this point. About three G's."

For just over one minute Schirra was silent and then continued with his narrative,

"Still in a relatively horizontal attitude. Rate command is working well. Glad she's holding. Doing very well. Coming up on five and one half (G's). Rate command still holding, fuel is still 70, seems low. Coming up on six and one half, seven G's. Coming up to eight G's. Re-command holding. Taking a pretty big yaw out. Not too bad, I have it pretty well. Manual (fuel) is 60 percent. She flying it very well. Coming off... peak-G was indicated at seven and one half."

Schirra's narrative was then interrupted by Hawaii CAPCOM Gus Grissom who established post-blackout communications. Dropping through the upper atmosphere, SIGMA 7's reaction control system kept the spacecraft stable. As the capsule dropped through 40,000 feet the drogue chute automatically deployed while at the same time Schirra punched the deployment button. After passing 20,000 feet the main chute deployed and shortly thereafter de-reefed. Schirra happily described his parachute saying.

"... looks like a sweetie pie."

Reentry was so close to his landing target that the crew of his recovery ship, the USS KEARSARGE, saw his capsule's contrail in the sky and heard its sonic booms.

(See Image 16, page 89.)

SIGMA 7 landed with 60 percent of its manual fuel remaining and 50 percent of its automatic fuel – an amazing demonstration of precision flying.

Some documentaries, made decades after the mission, would have you think that at the time of SIGMA 7's flight the mission went somewhat unnoticed by the public in general. Nothing could be farther from the truth. In fact, this mission was big news. It was widely covered worldwide and followed closely by every news outlet in the United States. The flight made headlines and Wally Schirra became an instant celebrity. Even in my kindergarten class, Wally's name was a part of our daily lessons. We knew who he was and we knew that he had flown in space, even though the concept of "spaceflight" was a bit beyond our world of Tinker Toys and nap time.

Right around the time of SIGMA 7's flight, however, our class got busy building a piece of aviation of our own: an in-classroom airliner! Using some huge wooden timbers that normally served as forts or log cabins, our teacher, Mrs. Cole, helped us form the outline of an aircraft's fuselage. Inside it we lined up the chairs from the classroom in the fashion of passenger seats and had two seats in the front for the "pilots." Remembering the early 1960s mentality, the girls took turns as the "stewardesses," and each day two of the boys got to wear ear muffs and play the parts of "the pilots." I could hardly wait for my turn to wear the headset ear muffs and be the pilot. Every day two other boys were picked to be the pilots, but by Friday, they would be out of choices and it would be my turn! I got to school that day and discovered that the airliner was now a fort! A lousy stinking fort; my airliner was gone! The least that they could have done would have been to make it Mercury Control, or something, so I could still wear the earmuffs headset, but NO, it was a fort! This could have been a great lesson in Project Mercury, instead it was F-Troop for toddlers. Mrs. Cole tried to pacify me by telling me that not everyone can be a pilot; it did not work. I wondered if the Mercury astronauts got to wear the ear muffs. Years later, as a professional pilot, every time I put on a headset I remembered my ear muff envy and that lousy stinking fort.

Interestingly, the official NASA press conference for Wally Schirra was not held at Cape Canaveral. Instead it became the first official NASA press conference held at the new home of the manned space program: Houston, Texas.

Wally Schirra went on to command Gemini 6 and fly the first rendezvous in space by orbiting in formation with Gemini 7. He later went on to command the first flight of an Apollo spacecraft, Apollo 7, just six years after his Mercury flight. Schirra was the only man to fly Mercury, Gemini and Apollo spacecrafts. He retired from NASA after the Apollo 7 mission, but remained a favorite of almost everyone who was involved in, or a follower of spaceflight. A big part of that popularity came from his co-hosting Apollo's 11 through 17 as well as the Skylab missions with CBS's legendary spaceflight anchorman Walter Cronkite.

SIGMA 7 demonstrated the Mercury spacecraft

could effectively fly missions that lasted longer than three orbits. Schirra also showed that proper handling of the suit cooling system could make the B.F. Goodrich IVA spacesuit practical for an extended spaceflight beyond seven orbits. Additionally, he proved that the "mouse fart" control firings worked quite well and greatly extended the spacecraft's fuel supply. At length, he also found that the astronauts could do without the Mercury spacecraft's periscope. That was important because the area occupied by the periscope was needed for an experiment that was to be carried on the next Mercury mission.

FAITH 7:
WHAT AN AFTERBURNER

Project Mercury was scheduled to conclude with the Mercury-Atlas 9 (MA-9) mission. Following Wally Schirra's successful flight and Deke Slayton's removal from active flying only one of the original seven Mercury astronauts had yet to fly, and that was Gordon Cooper.

From the point of view of a historian, especially one who had grown up with spaceflight, Gordon "Gordo" Cooper was probably the embodiment of what all of us who were looking at the space program from the outside imagined an astronaut to be. A good illustration of that can be found in Gene Kranz's book "Failure Is Not an Option." On page 14 of that terrific book Kranz describes his first encounter with an astronaut who just happened to be Gordon Cooper. Paraphrasing that text, Kranz says that he had landed at Patrick Air Force Base in Florida, which is just south of the Cape, and did not have a lot of sense as to exactly where he was. Kranz stepped onto the ramp and tried to get his bearings. It was then that he saw a "…shiny new Chevrolet convertible…" and an Air Force enlisted man get out and hold the door for a guy in civilian clothes. The man in the Corvette offered Kranz a ride to the Cape. It turned into the ride of his life as the driver blew through stop signs and broke every regulation in the book as they screamed up the road to Cape Canaveral. It turned out that the hot-rodder was Gordon Cooper.

That encounter took place on November 2, 1960. Those were the Wild West days of manned spaceflight and true cowboys like Gordo Cooper fit right in. Two and one half years later Cape Canaveral was still the Wild West of spaceflight and the Mercury Seven astronauts still ruled the roost. The only difference was that by May of 1963, only two of the original seven astronauts had yet to fly. Deke Slayton had been medically removed from flight status because every once in a while his heart skipped a beat. He had since been assigned the position of Chief Astronaut. Now Gordo Cooper was the last man in line for the last flight on the Mercury schedule. The reasons why Gordo was selected to be last in line to fly a Mercury flight are as shaded in as much secrecy as the reasons why Shepard, Grissom and Glenn were selected to be first three. NASA wanted to present the image of the astronauts being a team of equals with no one being better than the other. Yet each had their faults and each had their strengths, thus each ended up in their own private place in American history. Gordo's place would be to fly the first day-long mission in NASA history. It was an assignment that would play well to his strengths.

Atlas booster number 130D was scheduled to loft Cooper into orbit, but the requirements of a day-long mission as well as some glitches in the booster pushed back the launch date. In fact, Atlas 130D had to be sent back to its manufacturer in order to correct wiring defects. Once it returned to the Cape the standard small glitches that were common in the days of Mercury took their toll on the overall schedule and the launch date eventually slipped to May 14, 1963.

Another issue in the day-long flight was the fact that Cooper would need more consumables aboard his spacecraft in order to keep him alive for the greatly extended mission. The Mercury spacecraft originally had been designed for a three-orbit mission and Schirra's SIGMA 7 spacecraft had to be modified just to add another three orbits. Now Cooper was slated to fly as many as 22 orbits, so his spacecraft needed some weight scrubbing in order to slim it down and allow for the extra consumables. Removed from the MA-9 spacecraft were items such as the low level commutator, the periscope (which Wally Schirra had found to be almost useless anyway), the rate control system, the backup telemetry transmitter and the backup UHF voice transmitter. Added were nine pounds of cooling water, eight-tenths of a pound of carbon dioxide absorber, four-and-one-half pounds of drinking water, a urine transfer system, a parallel coolant control valve, four pounds of breathing oxygen, a slow scan TV transmitter and 15 pounds of additional fuel. In the end, Cooper's spacecraft wound up weighing three pounds more than Wally Schirra's.

FAITH 7:
BUZZING THE BOSS

Cooper himself was completely responsible for an event that took place just one day before his scheduled launch. The incident involved Cooper flying an F-102 around the Cape and "buzzing" buildings; in particular the building where NASA management had their offices. Years later Cooper talked about the incident, from his perspective, in his NASA oral history.

"That was the day before the launch, and we had a firm ground rule that we didn't put anything new within the spacecraft or into the pressure suit or anything at the last minute that you could not adequately test. But the medics had overridden this very firm ground rule we had, and they had cut a big hole in the side of my suit and inserted another great big metal fitting, which rubbed the heck out of my ribs, in order to put in the hose for the blood pressure cuff that could automatically inflate and they could take blood pressure whenever they wanted it. And this really antagonized me because we had broken all our ground rules here at the last minute. We also had a little ground rule that we would always go out the day before launch and get one of our fighters and take it out and really shake it down and shake out all of the kinks. So, being a little angry at the time when I went out to get my fighter ready to go, I came back over the Cape and buzzed the Cape fairly low; and when Walt Williams (the Deputy Director for Mission Requirements and Flight Operations) happened to walk out of his office and almost fell flat because I was pretty low coming right by his window, lighting the afterburner, he was a little startled. And he was pretty teed off about it, I think."

There are many accounts in the annals of history concerning project Mercury that try and describe Gordo's "buzzing incident." Some say that Walt Williams looked out of his office window and saw Cooper's F-102 streaking past below his level. Almost all of the tales concerning this event state that Cooper lit his afterburner right in front of the building in order to put an exclamation point on his anger. Of course Walt Williams was not about to just let the incident pass. He immediately phoned Deke Slayton, who was then chief astronaut, and demanded that Cooper be taken off the MA-9 mission. Slayton, later stated in his book, "Deke" that he suggested it would be better to just "sweat" Cooper for a while. Slayton then told Gordo that Williams was going to ground whoever had been buzzing the Cape, and then just left it at that. No doubt, Cooper sweated.

Of course this was not the first time that Gordo had gotten crosswise with NASA management. He had raised some eyebrows when he elected to name his spacecraft "FAITH 7." NASA management felt that if something should happen and they should lose the spacecraft, perhaps in a similar manner to the way that LIBERTY BELL 7 had been lost, the headlines would read, "NASA loses faith." The name, however, endured in spite of management's misgivings. In another case, Gordo made waves during the countdown demonstration test where a full dress rehearsal of the launch took place. As the astronaut transfer van pulled up to the pad, Cooper got out and marched confidently toward the gantry elevator in front of the world press. Then he stepped into the elevator, turned around, and acted as if he were frightened and wanted to turn back. The technicians played along and appeared to force him back into the elevator and up the gantry. Although everyone in the Cape news media knew it was a joke, it played better among Cape Canaveral insiders than it did in the public at large. I recall sitting on the front porch of our house late in the evening in the summer of 1963 with my dad, watching a satellite pass overhead. Dad quipped,

"Well, at least there's not some guy in there screaming 'Let me out.' "

I spent many years wondering who that astronaut was.

Of course, Gordon Cooper was not removed from the MA-9 mission. In the predawn hours of May 14, 1963, Gordo suited up and headed to Launch Complex 14. Soon he was happily crammed into his tiny capsule. Unlike his predecessors, the weather was not the cause for scrubbing the first launch attempt. Instead, a breakdown in radar systems at the Bermuda tracking station delayed the launch. Intermittent digital data in both the range and azimuth channels of the C- band radar at Bermuda became apparent. The countdown, however, continued on until the T-60 minute mark. At that point another unscheduled hold in the count was called. This delay was caused by the failure of a fuel pump in the diesel engine that was used to move the gantry's transfer table. Once that hold was cleared the countdown continued until the T-13 point. By that time the radar issue at Bermuda had not been cleared and the launch was scrubbed for the day. Cooper returned to hangar "S," where the astronaut quarters were located. The following day Cooper again went through the process of medical examination, suiting up, transfer to the pad and being loaded aboard the spacecraft. He completed his checklists and then took a little nap. As the countdown came out of its preplanned hold, Cooper was awakened and rapidly went back to work.

FAITH 7 lifted off just shortly after 8:04 a.m. CAPCOM for Cooper's launch was Wally Schirra and as Cooper lifted off he called out, "SIGMA 7, FAITH 7 on the way," acknowledging his fellow astronaut's call-sign. As Atlas 130D roared toward space, Cooper commented, "What an afterburner!" Indeed, this time Walt Williams had no problem with Gordo's afterburner.

At 00:01:45 into the boost Cooper suddenly experienced several large yaw oscillations of about plus-or-minus six degrees per second. Momentarily this concerned him as he felt that the automatic Abort Sensing and Implementation System (ASIS) would do its job and fire the escape tower, aborting the mission and landing Cooper a short distance down range. The oscillations, however, dampened out as the boost continued and soon FAITH 7 was flying smoothly once again.

Staging, or Booster Engine Cut-Off, where the Atlas drops its lower boat tail and its two outboard engines leaving only the center "sustainer" engine to continue the boost, was very noticeable to Cooper. He reported later that it began with a "glung" followed by a sharp crisp "thud." He found that although the sounds of BECO were very pronounced and prominent, there was almost no change in acceleration.

Escape tower jettison took place shortly after staging. It came a bit later than Cooper had anticipated and just as he was reaching for the handle to manually jettison the escape tower, it simply departed. He reported seeing it at about 100 yards out, off to his left and spiraling into the distance. The remainder of his boosted flight was very smooth and went exactly as he had expected. At Sustainer Engine Cut-Off he felt a rapid decrease in acceleration and then heard a series of sounds similar to BECO as the capsule separated from the booster. With separation three small posigrade rockets on the retro package fired pushing the capsule from the booster. Cooper reported that they, "… gave me a distinct boot in the rear."

FAITH 7:
A FLASHING LIGHT AND WATER,
WATER EVERYWHERE

As with previous Mercury astronauts, Cooper became somewhat enamored with the sunsets as seen from orbit: "I never tired of looking at sunsets. As the sun begins to get down towards the horizon it's very well defined, quite difficult to look at, and not diffused as when you look at it through the atmosphere. It is a very bright white; in fact, it is almost the bluish white color of arc. As it begins to impinge on the horizon line, it undergoes a spreading, or flattening effect. Sky begins to get quite dark and gives the impression of deep blackness. The light spreading out from the sun is a very bright orange color which moves out under a narrow band of bright blue that is always visible through the daylight period. As the sun begins to go down, it is replaced by this bright gold-or orange band which extends out for some distance on either side, defining horizon even more clearly. The sun goes below the horizon rapidly, and the orange band still persists then gets considerably faint as the black sky bounded by dark blue bands follows it on down. You do see the glow after the sun has set, although it is not ray-like. I could still tell exactly where the sun had set a number of seconds afterwards. The first indication I got of the sun coming up behind me was the lighting of the clouds from underneath. I noticed the clouds getting lighter and lighter, and I can still see the stars. Suddenly, my window would get into the oblique sunlight and appear to frost over just as an aircraft canopy does. Frost was apparently the result of a greasy coating on the inside of the outer pane which completely occluded my vision under the right conditions."

One of Cooper's major experiments during his day-long mission was to deploy a small softball-sized sub-satellite that had two flashing strobe lights attached. Known simply as, "the flashing beacon," this little device was deployed from the area of the spacecraft that had once contained the periscope. Cooper reported that at 03:25:00 mission elapsed time he went to fly-by-wire low, and slowly pitched the spacecraft 20 degrees nose down. He then pushed the button to deploy the flashing beacon. There was a loud, "cloomp" as the squib that held the beacon in place fired and the device was deployed.

"I never did see the beacon on that first night," Cooper reported later, "but I was having some difficulty finding my 180 degree point. I tried unsuccessfully to observe the flashing beacon early on the day side also."

Cooper had better luck on his next orbit.

"On the second night side after deploying the flashing beacon, shortly after going into the night side, I spotted the little rascal. It was quite visible and appeared to be only 8 to 10 miles away. I deliberately moved off target, waited until 5:40 (MET) and eased back to 180 degrees yaw and saw the light again, at which time it appeared to be around 12 to 14 miles away and still quite visible. On the third night side after deploying the flashing light I had no anticipation of seeing it all; but at 6:56 there it was, blinking away. It was very faint and appeared to be at a distance of about 16 to 17 miles. I would say it was approximately the brightness of a fifth magnitude star whereas on the second night side after deployment it appeared to be about that of a second magnitude star."

The flashing beacon experiment had been a success.

Of course, Cooper's mission was not without its problems. Not all of those problems were high-tech, in fact, some were as basic as the issues one might experience when going camping. About eight hours into the flight Cooper noticed that the valve on his drinking water container was leaking. In fact it was leaking so badly that he was unable to put any water into the plastic food containers in order to rehydrate his freeze-dried food. He discovered quite to his discontent that in zero-G there was no way to work any of the water from the nozzle through the container into the food. The container itself was part of the problem because its lid was so difficult to close that whenever he did get some water into it he was not able to work it into the food. The water simply came out of the top of the container. Cooper ended up with water and a little bit of food all over himself, his gloves and his instrument panel. At length, he simply gave up on it and spent the rest of the mission eating only the snack-type, bite-size foods.

Another small and frustrating problem that occurred took place with the "condensate bag." This "bag" was a part of the system they collected excess moisture from both Cooper's suit and spacecraft environmental control system. When the astronaut saw that the bag was filled, or close to being filled, he was supposed to pump that collected moisture into a 3.86 pound tank. The problem was that the condensate container filled up much faster than expected. When Cooper attempted to transfer the contents of the bag into the tank the pump appeared to jam. When he tried to switch back to "the other tank" the needle fitting came out of the entire system and became useless. Cooper's only choice was to begin pumping the condensed water from the bag into his regular drinking water tank. That, of course, contaminated his onboard water supply. He was thus relegated to go without drinking for most of the flight and when he was thirsty he had to drink from his survival kit's water supply.

FAITH 7:
STARTLING, WONDROUS THINGS

Some of Cooper's more interesting observations were those of the Earth itself as well as the moon.

"At night I could see lightning," he recalled, "Sometimes five or six different cumulus buildups were visible at once in the window. I could not see the lightning directly but the whole cumulus mass of clouds would light up. From space, ground lights twinkle, whereas stars don't twinkle. I could not distinguish features on the moon. It was a partial moon at night, but it was full when it was setting in the daytime. It was quite bright at night, but on a day side it was a lightish blue color."

(Insert Image 17, page 90.)

Yet another aspect of the day-long flight of FAITH 7 was that Gordon Cooper would become the first American to actually sleep in space. But "sleep" was something of a relative term. He encountered no difficulty in being able to sleep once the spacecraft was in drifting flight and completely powered down. He found the situation to be "relaxed, calm, and floating free." Although he slept quite soundly his sleeping was done in a sequence of naps. Some of his naps were as short as an hour while others were nearly three hours long. He found that he slept so soundly that on one occasion, after about a one-hour nap, he woke up feeling that he had no idea where he was and it took him several seconds to re-gather his senses. Similar awakenings took place after each nap, however he found that he was always able to easily regroup and regain his senses.

A few of Cooper's ground observations became quite controversial. After his flight Cooper claimed that he was able to "detect individual houses and streets" and "what appeared to be trains and trucks in some of the clear regions of the Tibetan area." He also said that he was able to observe wind direction and velocity on the ground from smoke emanating from smokestacks or fireplaces of houses. This was particularly true in the Himalaya area where Cooper claimed he could see, "a lot of snow on the ground in the upper portions of the mountains and a lot of lakes frozen over even down in the lower sections of the wind-blown sand of the high plateau areas." For a number of years following the flight of FAITH 7, scientists and engineers alike debated whether or not such details could actually be seen from orbit by the human eye. It was not until the beginning of the Gemini program that Cooper's fellow astronauts discovered that similar observations could actually be made.

On revolutions 14 through 20, FAITH 7 orbited beyond the range of most of the normal tracking stations and thus, for the most part, it was close to being completely out of communication with the ground. Fortunately things proceeded routinely during that entire portion of the flight. Just prior to the 22nd hour of his flight, while in a prolonged period of being out of contact with the ground, Gordon Cooper made mankind's first prayer in space. Speaking into his onboard tape recorder he prayed.

"I would like to take this time to say a little prayer for all the people, including myself, involved in this launch and this operation. Father, thank you, for the success we have had in flying this flight. Thank you for the privilege of being able to be in this position, to be in this wondrous place, seeing all these many startling, wondrous things that you've created. Help guide and direct all of us that we may shape our lives to be good, that we may be much better Christians, learn to help one another, to work with one another, rather than to fight. Help us to complete this mission successfully. Help us in our future space endeavors that we may show the world that a democracy really can compete, and still is able to do things in a big way, is able to do research, development, and can conduct various scientific, very technical programs in a complete peaceful environment. Be with all of our families. Give them guidance and encouragement, and let them know that everything will be okay. We asked this in thy name. Amen."

FAITH 7:
OTHER THAN THAT, THINGS ARE FINE

Trouble for Gordon Cooper came on his 19th orbit. While switching off cabin lights in order to become more dark-adapted and thus perform some scientific experiments, Cooper saw that his .05 G light had illuminated. That light normally comes on during the reentry when the spacecraft's accelerometers sense deceleration as a spacecraft begins to contact the Earth's atmosphere. Over the next two orbits, while engineers on the ground worked at troubleshooting the .05 G light, FAITH 7's ASCS inverter, a device for changing DC power into AC power, failed. This meant that Cooper's automatic controls were simply gone. When Cooper switched to the backup inverter, he found it, too, was inoperative. Soon, FAITH 7 came into contact with the Hawaii tracking station.

"Well," Cooper reported very calmly, "things are beginning to stack up a little. ASCS inverter is acting up, and my CO2 is building up in the suit. Partial pressure of 02 is decreasing in the cabin. Standby inverter won't come online. Other than that, things are fine."

Ground control advised Cooper to take a Dexedrine tablet to sharpen his awareness of his current situation. That didn't do anything for his inverters, however.

It was clear that in order to get back to the surface

of the Earth safely, Gordon Cooper was going to have exactly three navigation tools that he could count on: his wristwatch, his hands on the manual controls and his eyeballs out the window. Aboard the tracking ship COSTAL SENTRY QUÉBEC, astronaut John Glenn, serving as CAPCOM, read the reentry procedures up to Cooper and they started the countdown. At five seconds prior to retrofire, Cooper armed the retro squibs and at zero he punched the button. The retro rockets fired exactly on time. Using his view out the window and the manual controls on the spacecraft, Cooper held FAITH 7's attitude through the burning of the retro rocket sequence. Once the retros had finished firing Cooper manually jettisoned the package. Flying through the reentry itself Cooper felt the spacecraft tend to "wallow" in a lazy small spiral. This wallowing was of a very small magnitude that he was easily able to handle. Then between 95,000 feet of altitude and 50,000 feet the spacecraft began a series of increasing yaw oscillations. Below 50,000 feet the oscillations became quite large and difficult to manage. Those oscillations were muted at 42,000 feet by Cooper's deployment of the drogue chute. The main chute was deployed at 11,000 feet and Cooper was on his way to a safe landing.

Considering that Gordon Cooper had lost all automatic controls and had to fly with just the use of his eyes, window and control stick, he managed to bring FAITH 7 down just 1.21 miles from the recovery carrier the USS KEARSARGE. It was an amazing feat that clearly demonstrated that among millions of dollars worth of technology, the human element was a key factor in the success of this mission.

(See Image 18, page 91.)

MERCURY ENDS

Project Mercury was over and the United States' human spaceflight effort went into its first "gap" between programs. Oddly, had it not been for President Kennedy's decision to send Americans to the moon, the entire United States human spaceflight program could very well have dissolved and ended with FAITH 7's mission. Instead, the nation's public will, as well as political will was to push ahead, continuing toward placing Americans on the moon. The next program coming up was scheduled to fly in late 1964 but actually would not fly until early 1965. Originally that program was called, "Mercury Mark II" in part because the Mercury program had already jumped through all of the congressional funding hoops and a new program may require re-jumping, but also because a more advanced Mercury spacecraft had been in the works prior to the flight of FREEDOM 7. An extension of the Mercury Program, would thus negate a lot of red tape. Yet, ultimately, Mercury Mark II was renamed Project Gemini and work was well under way as Mercury ended.

But there was a very small group within NASA who wanted to keep the one-manned version of Project Mercury going; chief among them was Alan Shepard. Shepard felt something might be gained by sending him up into orbit aboard a Mercury spacecraft on an open-ended mission and allow him to simply orbit until, as he stated in a later interview, "...we ran out of just about everything." NASA management disagreed saying that Mercury was finished and now they wanted to press on with Gemini. NASA administrator James Webb was particularly dead set against another mission, especially an open-ended mission. The fact was Cooper had just done that. He had taken the Mercury spacecraft up and flown it until he ran out of "just about everything," so what would have been the point of another mission doing the same thing? Undeterred, Shepard took the opportunity of the White House ceremony meant to honor Cooper's flight to personally bring up the subject with President Kennedy. The President, however, sided with Administrator Webb, and Project Mercury officially came to an end. Shepard's MA-10 capsule now resides as a footnote display at the Udvar-Hazy branch of the National Air and Space Museum. To incurable space-buffs, such as me, it is kind of fun to go and see that vehicle and ponder whether Shepard, when he "ran out of everything," would have been able to land closer to the recovery carrier than Cooper. It would have been a "first against the last" contest in accuracy landing – an alpha, omega among the Mercury 7 astronauts.

For those of us who were growing up with the United States' manned spaceflight program, there would be much, much more to come after Mercury. Construction was under way in Houston, Texas as the new Manned Spacecraft Center was being completed. At Huntsville, Alabama, tanks for the huge Saturn V as well as other full sized components were being

tested. Near Edwards Air Force Base, the first production F-1 engine for the Saturn V's first stage had just
finished testing and was being readied for shipment
to Huntsville. Additionally at Huntsville, as well as
at the Mississippi Test Facility, giant test stands and
huge support complexes also were being constructed
to support the Saturn V. Through the southern states,
work boots and bulldozers were covered in red clay
as preparations to go to the moon were spooling up.
At Cape Canaveral and at the Merritt Island Launch
Area, which would later become the Kennedy Space
Center, new construction also was taking place.
Launch Complex 37 was being completed from
which the Saturn I Block II and the up-rated Saturn
I, later known as the Saturn IB would be launched.
Meanwhile Launch Complex 39, from which the
Saturn V would launch, also was under construction.
So although Project Mercury was over, positive steps
toward a future for America's manned space program
were under way, and millions of us would get to grow
up with it and be inspired by it. The next step would
be Gemini and it was just over a year and a half away.

MY STRANGE RELATIONSHIP
TO ENOS THE CHIMP

In some obscure ways, I have always felt as if I had some sort of a connection with the space program. For example, I am the same age as Enos the chimp. So, I guess that relates in a strange way. Of course, he was practicing for a flight onboard an Atlas rocket at Cape Canaveral at the age of three when I was still killing boxelder bugs with a little hammer on the side of our neighbor's garage. Yet I did not have to go through electro-shock training either. They saved that until I began flight training in 1978.

Now, if you came to read this chapter expecting something else about me and Enos, that is because I have cleverly suckered you into reading this section of the book, which is normally called something like "Author's notes" or "Postscript" or "About this Book" or my favorite, which NASA often uses: "How to Use this Book," or something boring such as that which would be correct by literary standards. Normally folks just skip that part, but there is important stuff in it. So I figured that I owe it to you, the reader, to cleverly lure you into reading this. You also get to learn a bit about me, the author, which will allow you to gain context for the chapters that follow.

Considering that I was a pre-schooler and

kindergarten kid during all of Project Mercury, although I was indeed growing up during the program, my own memories are quite scarce. Thus, in order to prevent this book from being just a dry re-hash of the early days of the program I solicited the memories of other folks who were older kids during that era. Each was invited to send me their story, no matter how small, and allowed me to publish it. Each was given the option of presenting their home town or general area of residence and for anonymity's sake I allowed them to give me a first name only to attach to their story. The story was presented as "(OTHER Memories)." These stories were taken at face value with no background research and were only edited for punctuation and grammar. None of the contributors wanted their full names published.

A few words about the "style" in which this book is written. I do my best writing in a "conversational style" because I want the book to talk to you the reader. I have always thought that the best way to teach history is through narrative and that requires that I tell you the story and take you to the place where the events have happened. Thus, I have asked my editors to consider that while editing the text. Certainly, my style is not firmly in line with the "AP Style Guide" or some such thing, but neither am I. So if, while reading the stories in this book, you begin to feel as if I am talking to you rather than you are reading a PhD. thesis, I have done the job that I set out to accomplish.

Some footnotes about footnotes… I hate them, thus I have never used them in any of my books. Yes, I know it is not proper and all of that rot, but this is not a technical report, nor a thesis for a Master's Degree.

This is a book for pleasure reading. Footnotes break up the flow of the narrative. You will find reference sources in the back of the book.

I managed to grow up with an insane eye on spaceflight, then I went to college, became a professional pilot and airline captain and after three furloughs in 11 years, decided to find another career. Hey- don't get me wrong, flying is a great job- you cannot beat the view, but there are so many people in that business who are determined to just suck all the fun out of it. Today, I make my living as a research historian and also manage to cover space stuff as the Spaceflight Analyst for the Aero-News Network. Each time I go to the Kennedy Space Center for a launch, get my press credentials and drive through the gate, I turn right back into that little kid who sat cross-legged in front of that old cabinet TV set and watched a launch with the same level of anticipation and thrill that a kid gets at Christmas. Standing at the press site I look at those buildings and can almost see Cronkite, Chancellor, McGee and Bergman sitting there broadcasting. This series of books is intended to bring some of that same feeling back to those of you who also grew up with spaceflight, but cannot get into the press site with me.

Additionally, this series of books is designed to present the perspective of those of us who, like me, actually grew up with the United States' manned space program. It is my hope that in many ways you will relate to the stories told here, as well as be informed by the details of the missions that I present.

For those of you who were born too late to witness the days of Mercury, Gemini, Apollo and the early Space Shuttle, let this be your personal time machine.

It is my hope that you will be taken back to the days of your parents, aunts and uncles, and even grandparents, so that you too may capture some of the essence of what it was like to grow up with spaceflight.

If you are a person who grew up with spaceflight while maintaining just a casual interest and you read this book and enjoy it, I will have done my job. If you are a hard core space buff and you read this book and learn just one single fact that you had not known before, I have also done my job. If you are one of those people who will gripe,

"What about the Russians? He hardly put any Russian stuff in the book!"

I'll say that the Soviets flew in total secrecy. If they wanted to be covered in historical narratives, they should not have flown in secret… so there.

SOURCES

"Gemini" Virgil "Gus"Grissom

"Lost Spacecraft- the Search for Liberty Bell 7" Curt Newport

"The History of Manned Spaceflight" Dr. David Baker

"The Rocket" Dr. David Baker

"History of Rocketry and Space Travel" Wernher von Braun and Fredrick I. Ordway III

"The First Small Step" Peterson's Book of Man in Space, Volume One

"The Unbroken Chain" Guenter Wendt and Russell Still

"Rocketman, Astronaut Pete Conrad's Incredible Ride to the Moon and Beyond" Nancy Conrad and Howard A. Klausner

"We Seven" by the Mercury Astronauts themselves

"US Spacesuits" Kenneth S. Thompson and Harold J. McMann

"Flights of the Astronauts" William Roy Shelton

"Man Alive in Outer Space" Henry B. Lent

"Exploring By Astronaut: the Story of Project Mercury" Franklyn M. Branley

"Countdown, the Story of Cape Canaveral" William Roy Shelton

"Americans in Space" John Dille

"Space Volunteers" Terence Kay

"Rendezvous in Space" Martin Caidin

"Answers to the Space Flight Challenge" Frank Tinsley

"Americans into Orbit" Gene Gurney

"Failure is not an Option" Gene Kranz

"Live from Cape Canaveral" Jay Barbree

"Opening the Space Frontier" Ray Spangenburg and Diane Moser

"This New Ocean" William E. Burrows

"Space Toys of the 60s" James H. Gillam

"Space Toys" Crystal and Leland Payton

"FREEDOM 7 Mission Reports" Robert Godwin

"FRIENDSHIP 7 Mission Reports" Robert Godwin

"AURORA 7 Mission Reports" Robert Godwin

"SIGMA 7 Mission Reports" Robert Godwin

Stratcat dot com; article by Gregory Kennedy "Stratospheric Balloons History and Present of their use in the Fields of Science, Military and Aerospace"

Video; "Man on the Moon" BBC (author's collection)

Video; "The light of Liberty Bell" NBC (author's collection)

NASA Johnson Space Center Oral History Project Oral History Transcript, Wayne E. Koons, Interviewed by Rebecca Wright, Houston, Texas – 14 October 2004

Oral History Transcript Alan B. Shepard, Jr. Interviewed by Roy Neal, Pebble Beach, Florida – 20 February 1998

Oral History Transcript John H. Glenn, Jr. Interviewed by Sheree Scarborough, Houston, Texas – 25 August 1997

Oral History Transcript M. Scott Carpenter Interviewed by Michelle Kelly, Houston, Texas – 30 March 1998

Oral History 2 Transcript M. Scott Carpenter Interviewed by Roy Neal, Vail, Colorado – 27 January 1999

Oral History Transcript Walter M. Schirra, Jr. Interviewed by Roy Neal, San Diego, California – 1 December 1998

Oral History Transcript L. Gordon Cooper, Jr., Interviewed by Roy Neal, Jet Propulsion Laboratory Pasadena, California – 21 May 1998

NASA Johnson Space Center Oral History Project Oral History Transcript Dee O'hara, Interviewed by Rebecca Wright, Mountain View, California – 23 April 2002

Oral History Transcript Maxime A. Faget, Interviewed by Jim Slade, Houston, Texas – 18&19 June 1997

Oral History 2 Transcript Maxime A. Faget, Interviewed by Carol Butler, Houston, Texas – 19 August 1998

Oral History Transcript Joe W. Schmitt Interviewed by Michelle T. Buchanan with Steven C. Spencer, July 1997, Friendswood, TX

 Post-launch Report for Mercury Atlas 1, NASA TR-X-56-746, August 2, 1960

Technical Information Summary, Mercury-Atlas Mission 3, NASA TM-X- 51300, April 17, 1961

Technical Information Summary, Mercury-Atlas Mission 4/8A, NASA TM-X- 51331, July 21, 1961

Technical Information Summary, Mercury-Atlas Mission 5, NASA TM-X- 51368

Technical Information Summary, Mercury-Atlas Mission 5/9, NASA TM-X- 51302, October 17, 1961

Post-launch Memorandum for Mercury Atlas 5, NASA N73-74161, December 6, 1961

Post-launch Memorandum for Mercury Atlas 6, NASA (un-numbered copy), March 5, 1962

Post-launch Memorandum for Mercury Atlas 9, NASA N75-76100, June 24, 1963

First U.S. Manned Six-Pass Orbital Mission (Mercury-Atlas 8, Spacecraft 16) NASA, TN D 4807, September, 1968

First U.S. Manned Three -Pass Orbital Mission (Mercury-Atlas 6, Spacecraft 13) NASA, TM X 563-I, March, 1964

Mercury Atlas 6 at a glance, News Release 62-8, NASA, January 21, 1962

Mercury Atlas 9 at a glance, News Release 63-90, NASA, May 19, 1963

Mercury Atlas 9 at a glance, News Release 63-90, NASA, May 19, 1963

Project Mercury Familiarization Manual, NASA, Manned Satellite Capsule, McDonnell Aircraft Corp., December 15, 1959

Project Mercury Maintenance Manual, NASA, Manned Satellite Capsule, McDonnell Aircraft Corp. December 15, 1959

Performance Characteristics of the Little Joe Launch Vehicle, Technical Memorandum, NASA, TM-X-561, MSC, Ronald Kolenkiewicz and John

C. O'Loughlin, September 1962

Data Analysis Report, Mercury Atlas 5, Spacecraft 9, NASA CR-52645, March 6, 1962

Space World Magazine, "The Human Factor" John Rublowsky, June, 1961

Space World Magazine, "The 40 Billion Dollar Question" Otto O. Binder, October, 1961

Space World Magazine, "How We Trained for Orbital Flight" Donald K. Slayton, March, 1962

Space World Magazine, "America in Space" Martin Caidin, May, 1962

Space World Magazine, "The Next Step 18 Orbits Around the Earth" Alan Gore, July, 1962

Space World Magazine, "The Frog Men Who Save Our Astronauts" James D. Rogers, July, 1962

Space World magazine, "Space Suits" James V. Correale and Walter W. Guy, September-October 1963

Space World Magazine, "Cryogenics...Products for the Realm of Supercold" November-December 1963

Space World Magazine, "Mercury Experience Applied" Jerome B. Hammack and Walter J. Kapryan, February, 1964

AUTHOR ACKNOWLEDGEMENTS

This is normally the section where I deliver thanks to libraries, researchers and private collectors, etc. Most of those who need acknowledgement will actually be found in the text or bibliography, so here I am going to place thanks upon those who are not in the text. There are a few ultra-important folks without whom I would never have produced this series of e-books. When I first decided to stick my toe into the e-book world, I did some forum reading and saw how many newbie e-authors were bashing their brains over one critical area: formatting. So I contacted my long-time friend, fellow professional pilot and author as well as ERAU alumnus, Mark Berry. He had already done some e-books and when asked he put me in contact with Kristina Blank Makansi, the editor and publisher at Blank Slate Press (blankslatepress.com) and founder of Blank Slate Author Services. With her guidance I discovered just how sheltered my life has been while working in print and how much my print publisher has done that I took for granted. She taught me about things such as proper photo DPI, bar codes, ISBNs and countless other areas upon which I would have stumbled. Joining Kristina in the effort was her daughter Elena who took on the task of formatting the interior and placing photos in the book. Together they form a professional team that was instrumental in bringing this book and those that will follow, into reality. I thank them. Finally I'd like to acknowledge Emily Carney, who owns and operates the FaceBook "Space Hipsters" site and a darned good spaceflight writer as well. She did not have a hand in making this book, but she really likes it when authors acknowledge her. And I'm in hope that this acknowledgement will inspire her to take a selfie with this book next to her ear and post it on her site. Thus, although she may not have had a hand in making the book, she will certainly have a hand in selling it.

Author Wes Oleszewski was born and raised in mid-Michigan and spent most of his life with an eye turned toward the space, flight and spaceflight. Since 1990 he has authored eighteen books on the subject of Great Lakes maritime history. Now he has turned his attention toward spaceflight.

Noted for his meticulous research, Oleszewski has a knack for weeding out the greatest of details from the most obscure events and then weaving those facts into the historical narratives which are his stories. His tales of actual events are real enough to thrill any reader while every story is technically correct and highly educational. Oleszewski feels that the only way to teach history in this age of computer and video games is through "narrative." The final product of his efforts are captivating books that can be comfortably read and enjoyed by everyone from the eldest grandmother to the grade-school kid and future pilot or historian.

Born on the east side of Saginaw, Michigan in 1957, Wes Oleszewski attended public school in that city through grade nine, when his family moved to the town of Freeland, Michigan. In 1976 he graduated from Freeland High School and a year later entered the Embry-Riddle Aeronautical University in

Daytona, Florida. Working his way through college by way of his own earned income, Oleszewski graduated in 1987 with a commercial pilot's certificate, "multi-engine and instrument airplane" ratings as well as a B.S. Degree in Aeronautical Science. He has pursued a career as a professional pilot as well as one as an author. He holds an A.T.P. certificate and has twice been elevated to the position of Captain. To date, he has logged three logbooks of flight time most of which is in airline category and jet aircraft. Recently he gave up the life of a professional aviator and now enjoys his job as a professional writer.

For more excitement, visit Wes's websites:
www.klydemorris.com
www.gwsbooks.blogspot.com

Made in the USA
Middletown, DE
05 March 2016